WHY REBEL

ABOUT THE AUTHOR

Jay Griffiths is the author of many books including *Wild: An Elemental Journey*; *Kith: The Riddle of the Childscape*; *Tristimania* and *Pip Pip: A Sideways Look at Time*. She won the Discover award for the best first-time author in the USA; the inaugural Orion award and the Hay International Fellowship. She has broadcast and written widely, including for Radiohead and the Royal Shakespeare Company. Her work has received widespread accolades including from Gary Snyder, Barry Lopez, Don Paterson, John Berger, Philip Pullman, KT Tunstall and Nikolai Fraiture.

WHY REBEL

Jay Griffiths

PENGUIN BOOKS

PENGUIN BOOKS

UK | USA | Canada | Ireland | Australia
India | New Zealand | South Africa

Penguin Books is part of the Penguin Random House group of companies
whose addresses can be found at global.penguinrandomhouse.com.

First published 2021
001

Set in 11.5/14 pt Fournier MT Std
Typeset by Jouve (UK), Milton Keynes
Printed and bound in Great Britain by Clays Ltd, Elcograf S.p.A.

The authorized representative in the EEA is Penguin Random House Ireland,
Morrison Chambers, 32 Nassau Street, Dublin D02 YH68

A CIP catalogue record for this book is available from the British Library

ISBN: 978–0–241–99272–2

www.greenpenguin.co.uk

Contents

Introducing

I wish that everyone who said they believed in angels would actually believe in insects. Thanks to the insects, we have food, but it is also insects who, tactfully and quietly, remove the dead. Without them, we would be wading through corpses at every step.

We need the living world, that truest of worlds, both in death and in life: thanks to other animals, human life is vitalized. Animals, their senses alert to every scent and whisper through their paws, antennae, wing-tips and noses, are our guides and our healers. Mindedness surrounds us: in air and water, on land and in the soil under our feet where the lovely, holy worm, in sweet complicity with the dreaming Earth, quickens death itself to life. In the transforming dark, trees are messaging each other through their roots and, where tree roots link with fungi, the network of underground awareness is able to respond differently to the presence of different creatures. Awareness surrounds us. Tingling.

Perhaps nowhere is so full of life as the forest of the Amazon: bursting, writhing, tickling and rotting. Death itself is merely matter requickened into something else because everything is a part of the shapeshifting metamorphosis at the green core of life. In the Amazon I felt, as never before, that I had touched the quick of the thing, had put my finger to the pulse at the heart of the world, in an experience of empathy with that wildest of creatures, the jaguar. Indigenous societies in the Amazon have never forgotten the mindedness of life, honouring it with a way of thinking that is empathetic, willing to cross borders, linking with other species, a way of walking immersed in the true world, sensitive to its ethic.

How we walk the Earth has never mattered more than now. How we treat it, in the spirit of gift or theft, a bill or a prayer, has never been more important, not only for the intrinsic value of the living world, but for humanity's health and well-being. In a startling demonstration, the Coronavirus that has so traumatized the world was unleashed by our destruction of forest habitat and the slaughter of wild creatures.

Where were you when the music stopped? Everything ceased. Schools closed. Churches were shut. The pubs were silent. Friends could not meet. We cancelled everything to fight the virus: weddings, work, exams, football and theatre. We stopped doing everything except the very thing that had caused the virus in the first place: annihilating the living world. Forests have been destroyed for cattle to feed the insatiable appetite for beef: wildlife is slaughtered for tamelife as soya is grown on previously forested land to feed the chicken industry. Bereft of their seclusion in the forests, wild creatures were forced into contact and the virus jumped to humans. Astonishingly, the cause was quietly noted, and yet nothing was done.

Although humans are just 0.01 per cent of all life, we have destroyed 83 per cent of wild mammals, and we have done so without awareness. Now, aghast, we begin to see and to name. The saddest word I know in any language is the word *endling*. It refers to the last individual of a species, final and forlorn, at whose death the species is extinct. An endling is the very epitome of tragedy, and the word was coined, appropriately, in this age of extinction, this strange age of our dawning knowing, seeing the unhallowed horror that is approaching.

The human conscience needs to be corroborate with the natural world and aligned to its ethic, furthering a politics of kindness. But a political stance that is the opposite of kindness is on the rise: libertarian fascism with its triumphal brutalism, its racism and misogyny, the politics that loathes the living world. Tonally fascistic in many countries, it is openly so in some countries and, in the case of Brazil, where it launches its assault on the Amazon and the peoples of the Earth who live within those forests, it is a chiaroscuro of cruelty.

Here, then, the causes for rebellion: survival and awe; beauty and necessity; grace and grief. There is an uprising of life in rebellion for life, those who are griefstruck and furious for the tawny ones, the creatures of feather and fur, demanding that media and governments tell the truth about the emergency we are in, fighting for life in this shared, wild home. I have felt the grief, once diving in a coral reef at its dying moment, as it glows a luminous ringing blue, ultramarine, ultra-dying at an ultra-twilight. And I have been imprisoned in the cells for my activism for climate and against extinction. I write about my arrest and trial, and the judge's response. (In tears. 'Thank you. You *have* to succeed.')

Only when it is dark enough can you see the stars, and they are lining up now to write rebellion across the skies. Why rebel? Because nature is not a hobby. It is the life on which we depend.

The Solar Flares of Fascism

Fascism begins as something in the air. Stealthy as smoke in the darkness, easier to smell than to see. Fascism sets out an ethos, not a set of policies; it appeals to emotion, not fact. It begins as a pose, often a deceptive one. It likes propaganda, dislikes truth and invests heavily in performance. Untroubled by its own incoherence, it is anti-intellectual and yet it is contemptuous of the populace, even as it exploits the crowd mentality. Fascism is accented differently in different countries and uses the materials – and the media – of the times. It is hostile to egalitarianism and loathes liberalism. It champions 'might is right', a Darwinian survival of the nastiest, and detests vulnerability: the sight of weakness brings out the jackboot in the fascist mind, which then blames the victim for encouraging the kick. Fascism not only promotes violence but relishes it, viscerally so. It cherishes audacity and bravado, promotes charismatic leaders, demagogues and 'Strong Men', and seeks to flood or control the media. Even as it pretends to speak for the people, it creates the rule of the elite, a cult of violent chauvinism and a nationalism that serves racism.

The fascism of a Thomas Mair (who killed the MP Jo Cox) or the now-proscribed neo-Nazi National Action is so obvious that you can see it coming a mile away. The more insidious kind is the type being nurtured in today's Libertarian movement. Its precursors are in Italy, not Germany, in Italian Futurism, which bolstered Mussolini, in the poet D'Annunzio and in the mythic Roman figure of Deus Sol Invictus.

Filippo Marinetti, the poster-boy of Italian Futurism, articulated the emotional fascism from which political fascism stems. 'Our hearts are not in the least tired. For they are nourished by fire, hatred and speed!' Steel was the archetypal material for Futurist sculpture, but there are materials of the mind, too: the steel of cruelty, the gunmetal of hatred. 'We want to exalt aggressive action, the racing foot, the fatal leap, the smack and the punch.'

In contemporary Libertarianism there is a similar love of hatred. The Libertarian and far-right website *Breitbart* proudly published *In Praise of 'Hate Speech'* by Libertarian James Delingpole, while Sean Gabb, as Director of the Libertarian Alliance in 2006, said: 'We believe in the right to promote hatred by any means that do not fall within the Common Law definition of assault.' (Gabb said this as he stepped forward to defend David Irving's expression of Holocaust denialism as an act of freedom of expression.) The character traits applauded by today's Libertarians – ambition, *superbia*, speed, drive, spin, success and spikiness – are the qualities the Futurists valued. There is fire here, but never warmth; appetite, but never food.

Libertarians' bullyboy mentality detests the sensibility of liberalism, and torments those they call SJWs (Social Justice Warriors). There should be no regulations to protect the weak, they say, and they loathe the vulnerable: Milo Yiannopoulos, *Breitbart*'s star writer, having encouraged the racist and sexist abuse of actress Leslie Jones on Twitter then mocked her for 'playing the victim'. Libertarians scorn individual or cultural sensitivity in promulgating the term 'Generation Snowflake' to describe people who might 'melt' in the heat of hate speech or who want 'trigger alerts' to be issued over material that might traumatize survivors of sexual abuse. This attitude is proto-fascistic, to despise the victim for being vulnerable, using that weakness as a reason to treat them with contempt.

When the Libertarian Steve Bannon became Donald Trump's chief strategist, his appointment was cheered by the former head of the KKK, and approved by the American Nazi Party. Bannon collaborated with publishers of neo-Nazi White nationalist websites to turn *Breitbart* into a platform for the Alt-right, and became its CEO. Bannon describes Tommy Robinson, racist Islamophobe and former English Defence League leader, as 'the backbone of Britain'. He fosters links with the far-right, including Hungary's Viktor Orbán and France's Marine Le Pen, in an attempt to unite 'the

Judeo-Christian West'. Part of the playbook of Libertarianism in the USA and in the UK is Islamophobia. He has called for the beheading of Dr Anthony Fauci, the most senior US infectious diseases expert.

It was Libertarianism and the Alt-right that pushed Trump's success. His own ideology is too incoherent to fit exactly with any particular political grouping, but Libertarianism marks both his attitudes and those of his supporters, while his Alt-right allies seek the fascistic outcomes he supplies: his infamous refusal to condemn White supremacism at Charlottesville, for example, or his decision to use unidentified Federal Agents to attack legitimate anti-racist protesters (who Trump calls 'far-left fascists') at the Black Lives Matter protests in Portland. His campaign material included an insignia of an inverted red triangle, which was used in Nazi concentration camps to identify political prisoners.

Not all Libertarianism is proto-fascistic. Mainstream Libertarians claim an anti-statist position, whereas pro-statism is an almost obligatory part of a historic fascist credo. However, when anti-statism is used as a rationale to oppose all regulations, the most powerful are 'liberated' to exercise their power against the most vulnerable, and that is the emotional atmosphere in which fascism flourishes. It is about timbre, about the resonance between present political posturing and future political fact. Libertarians claim they hold an anti-authoritarian view even as they support the 'Strong Men'. I'm not saying that every member of the Alt-right, or anyone named here as espousing Libertarian views, is a fascist. I am, though, illustrating how closely one set of ideas (of Italian Futurism) maps with the matrix of ideas proferred by Libertarianism, and showing that fascism is almost ineluctably linked to Futurism.

The Italian Futurists, opposing political and artistic tradition, saw themselves as anti-establishment, as do contemporary Libertarians. They were driven, as the name suggests, forward to the future. As Marinetti wrote in the *Manifesto of Futurism*: 'Time and Space died yesterday. We already live in the absolute.' Libertarians, like the Futurists, loathe the past, which they associate with the natural world: the future is artificial, and they want to own it. Silicon Valley venture capitalist and Trump backer Peter Thiel, who describes himself as 'way libertarian', is heavily involved in The Technological Singularity, often referred to as The Singularity, promoting artificial superintelligence to create the end of natural history. Futurism wanted to create 'a new type of man'.

The Futurist Luigi Russolo championed a materialist idea of music as pure noise: sound should be 'bruitistic', exemplified in the noise of technology and the city, as opposed to the music of history or the sound of nature. He sought to brutalize the ear with noise. Language was reduced to shout, the hate-scream that features so heavily in contemporary Libertarian discourse. In design,

Libertarians go for the stark, crude smack of visual noise, not the sensitive subtlety of illustration.

The Futurists sought a language 'purified' by removing grammar. They created the poetry of noise, words released from the chains of grammar; unchain them, and pure sound could be set free. What is lost? Meaning itself. For grammar's necessary role is to shape and sculpt significance, to proliferate distinction and difference in the plurality of endless and beautifully diverse meaning: to the Futurists, though, grammar was merely a hindrance to a brutal gesture of noise. Trump's tweets brutalize language as well as people.

Italian Futurists (like modern Libertarians) had a special relationship to truth. Rumour and rhetoric was their register of choice. Futurist 'evenings' would include shouting political rhetoric at the audience to induce riots. Trump's press office produced the infamous term 'alternative facts', and in this post-truth era, lies are core to the Libertarian mission. Hannah Arendt, in *The Origins of Totalitarianism*, wrote: 'Before mass leaders seize the power to fit reality to their lies, their propaganda is marked by its extreme contempt for facts as such, for in their opinion fact depends entirely on the power of the man who can fabricate it.'

To lie is to abuse power. It is that simple. You know you're lying, and you also know that your audience doesn't know, so to lie is to exploit that power imbalance contemptuously. Boris Johnson's lies (the Brexit bus, the alleged EU rules against bendy bananas) illustrate his mocking disdain for the public. Trump's public lies are now counted in the thousands.

The Alt-right has a host of Alt-facts at its fingertips. Politics has always been riddled with propaganda, spin and cover-ups; the difference is that the Libertarian mind-set relishes its dishonesty: part trick, part game, part combat. Over and over, Libertarians use the notion of 'free speech' to vindicate outright lies (and offensiveness). Seeking freedom in all things, loathing restraint, the fact of lying becomes an ambition for breaking free of the tethers of honesty, of being 'bound' to tell the truth. Libertarians are masters of post-truth misinformation, and their technique is simple. Take a truth (climate change, Rwandan genocide) and simply declare it is a lie. (A 'myth' a 'hoax' a 'scam' a 'swindle'.) Sit back and wait a minute, and a very gullible press with little editorial responsibility will happily help you tell the public that they have been conned. No one wants to feel they've been duped so, hey presto! You have a 'war of ideas'. Without evidence, argument or proof; without expertise or knowledge.

Like any good fascist, Marinetti called for the burning of libraries, museums and academies. Fascism has never been known for its intellectuality and, although book burning seems a thing of the past, its equivalent is happening right now in the dissemination of lies about climate change, in flaming the scientists online, and trolling writers who cover the subject. Joris-Karl Huysmans, in 1884, noted how the supposed 'free thinkers' were really 'people who claimed every liberty that they might stifle the opinions of others'. In taking a deliberately anti-intellectual stance, Libertarians undermine the very idea of expertise, creating the thuggish intellectual atmosphere in which fascism flourishes. James Delingpole, discussing the rise of Trump, remarked: 'part of the problem is people with degrees: there are too many of them'. Libertarian Claire Fox promotes climate denialism, supporting Matt Ridley's attempts to undermine the science of climate change. Ridley's scientific area of expertise is what, exactly? Sexual selection in pheasants.

Central to the Futurist Manifesto was an adoration of the machine, to the point where the ultimate aim was the technological triumph of humanity over nature. Marinetti foresaw – and was intoxicated by – the idea of a war between organic nature and mechanized humanity. The Earth, the first and fundamental home of all, was loathed. Futurists fetishized cars, planes and technology in general, loving steel and detesting wood, which came gentle from the natural earth. They wanted to force the Danube to run in a straight line at 300km an hour, hating the river in its natural state ('The opaque Danube under its muddy tunic, its attention turned on its inner life full of fat libidinous fecund fish.')

Such detestation of the natural world, amounting to biophobia, is one of the hallmarks of Libertarians. Climate denialist James Delingpole (veteran of both *Breitbart* and its UK cousin *Spiked*) described environmentalists as 'scum-sucking slime creatures' and 'mutant slugs' in his Donald Trump Victory podcast and urged: 'smite them, salt them, and crush them underfoot'. Libertarians revile sustainability and object to environmental protection laws. Their contempt for the green movement was evident in Libertarian Martin Durkin's infamously deceitful prime-time 1997 Channel 4 TV series *Against Nature*, made by his company Kugelblitz, which shared the name of Nazi-manufactured weaponry.

That Durkin's title echoed Joris-Karl Huysmans' iconic Decadent novel *Against Nature* (1884) is telling. Huysmans' novel reads like a Libertarian tract in wanting to seek freedom from the unbearable restraints of nature. 'Nature,' Huysmans writes, 'has had her day . . . After all, what platitudinous limitations she imposes . . . what petty-minded restrictions . . . the moment has come to replace her, as far as that can be achieved, with artifice.' Artifice was valued by the Decadent movement, while the natural world, so beloved of the Romantics (and, interestingly, the Nazis) was loathed. Artifice is adored by Libertarians and the Alt-right. The Singularity is a Libertarian project. If Aldo Leopold suggested we could 'think like a mountain', then The Singularity's response is 'think like a machine'. In their dream of technological Singularity, humans become mechanistic. The arch-Libertarian Milo Yiannopoulos praises one Singularity-promoting blog (*LessWrong*), because it 'urged its community members to think like machines rather than humans. Contributors were encouraged to strip away . . . concern for other people's feelings, and any other inhibitors to rational thought.'

Kindness, concern and humanity, core to the best of human nature, are derided. In a pandemic such as Covid-19, those are the values that are needed. But the countries that have suffered the most (USA, Brazil and the UK) had leaders who were capable only of the opposite qualities. Where humility was needed, to heed advice and to act with care and caution, they displayed reckless *superbia*. Where compassion was required to protect the vulnerable, they responded with cruelty and carelessness. Where a respect for limitations was necessary (wearing masks and obeying lockdown rules), they showed their contemptuous Libertarian desire to be free of such fetters.

The refusal to wear facemasks was politicized. This was far more than a detail of partisan identity: it was an eloquent and significant expression of Libertarianism's creep. The facemask doesn't protect the wearer but protects others: it is inherently public-spirited, its ethos is communal. It is an expression of considerateness, particularly for the vulnerable. At the centre of Libertarianism is contempt for the weak: they recognize no victims, only people who 'play' victim. Libertarians hate regulations that protect others, be it environmental laws or facemask regulations. Libertarians loathe fetters on personal freedom: they perceive the facemask as a muzzle. As Libertarians promote the idea of 'free speech' at any price, and enjoy the practice of hate speech, as they campaign against all libel laws and 'cancel culture', so a masked mouth, covered and silenced, is the emblem of what they most detest: being gagged. So much so that the cover image of Claire Fox's book *I Find That Offensive* is a face with the mouth masked. As an extra twist, Covid-19 facemasks force a Libertarian to look a little like their enemy: Muslims wearing face coverings.

The aesthetic of Futurism and of Libertarianism follows the pattern of these emotional values. Steely, spiky and brutal. The statue that Trump commissioned of himself at Tulsa was in that brutalist style. The natural world, with its curves and gentleness, has no place in that mindset of engines, flight and rockets. Trump makes explicit his contempt for the living world – mere Earth – in his refusal to support environmental regulations, in his climate denial and in his seeking the unbridled technological artifice of colonizing space. Flight is at the heart of it. The flight from Earth to space: the Libertarian off-ground ideology of unlimited, unrestricted freedom. Fighting the fetters of carbon-emissions limits, the 'right to fly' was part of the 'freedom' the Libertarian ideologues sought, echoing Marinetti ('Hurrah! No more contact with the vile earth!'). The Futurists were obsessed with the rhetoric of flight, and the Futurist painter Giacomo Balla even named his daughter

'Propeller'. Aeropainting was a major expression of Futurism. One Futurist manifesto speaks mockingly of 'a terrestrial perspective'; in contrast, painting from an aerial perspective 'requires a profound contempt for detail'. (The contemptible details of my terrestrial perspective might include butterfly, pine marten and potato; grace, ice and plurality.)

When the British novelist Rex Warner wanted to expose fascism's vicious hatred of nature and of humanity, while also addressing its glamour, it was the image of flight he chose, and the rhetoric of freedom akin to the 'liberty' sought by Libertarians. In his 1941 novel *The Aerodrome*, the aim of the Aerodrome is to create 'a new and more adequate race of men' – handsome, cold, energetic, charismatic, cruel, rhetorically skilled, inhuman, cynical and fiercely anti-natural. The commanding fascist Air Vice-Marshal calls to the young airman: 'Your purpose – to escape the bondage of time . . . to obtain mastery . . . over your environment . . . This discipline has one aim, the acquisition of power, and by power freedom.' Flight is the way to escape the dank and oozing Earth, where the huddled unhygienic trudge their dumb ways. Above, superior, invincible and superb, is the airman, the Übermensch.

Libertarians have long loathed Indigenous cultures: the Libertarian cult figure Ayn Rand defended the genocide of Native Americans, saying that Native Americans did not 'have any right to live in a country merely because they were born here and acted and lived like savages'. In Brazil, the far-right president Jair Bolsonaro admires dictators including Pinochet. Nationalist, homophobic, misogynist, he is also dangerously racist towards Indigenous peoples. 'It's a shame that the Brazilian cavalry hasn't been as efficient as the Americans, who exterminated the Indians,' he has said. Bolsonaro has vetoed Covid-19 aid for Indigenous people, saying such provisions are 'against the public interest'. Due to Covid-19, Brazil lost a generation of Indigenous leaders and elders, and with their deaths went the tribal culture and medicine they knew. *We are facing extermination*, they have said. Meanwhile, wanting to end Indigenous rights to their territories, Bolsonaro has encouraged land raids to soar and has spoken of giving the ranchers guns. He appointed Ricardo Lopes Dias to head the department for voluntarily isolated Indigenous peoples, some hiding after massacres and epidemics. Dias, a fundamentalist missionary, advocates forced contact and open racism. He has worked for the New Tribes Mission, which, in the 1970s and 1980s, organized manhunts to capture tribal people, taking them to missions as unpaid servants and exposing them to diseases from which many died. The NTM (now renamed Ethnos 360) went fundraising recently for a new helicopter specifically to target tribes in isolation otherwise impossible to reach. Invincible from the air, the helicopter comes like a metal bird of prey. On a manhunt again.

Libertarians seek freedom to oppress people too powerless to stand up to them. They seek small freedoms, the freedom to travel with Covid-19 halfway across the country, as did Dominic Cummings, a man publicly endorsed by Steve Bannon, who called Cummings a 'brilliant guy'. They seek large freedoms: the freedom from the unbearable restraint of laws protecting the living world. The freedom to deny climate change when the effects of such denials are mass murder.

Climate-change denialism is the signature deceit of Libertarian rhetoric. Trump calls climate change 'bullshit' and 'a hoax'. His Environmental Protection Agency appointee Myron Ebell (who led a coalition of climate denialists to oppose the Paris Agreement on climate change) speaks of the 'myths of global warming'. In 2015, he referred to Pope Francis's encyclical on climate change as 'scientifically ill informed . . . intellectually incoherent . . . morally obtuse . . . theologically suspect . . . leftist drivel'. A year later, Delingpole (who doubts the extent to which climate change is man-made or catastrophic) writes in *Breitbart*: 'The global warming industry has been almost wholly under the control of crooks, liars, troughers, and scumbags . . . climate change is the biggest scam in the history of the world.' Another voice now: 'I made the mistake of thinking the truth would be its own ambassador,' says a climate scientist in a play by Steve Waters. If only.

'We will glorify war – the world's only hygiene,' bawled the Futurist Manifesto, in language that Serbia would recall in its ethnic 'cleansing' for the sake of militant nationalism. In the Yugoslav war, Bosnian Serbs were supported by a splinter sect of people in the UK, the Revolutionary Communist Party (RCP), who gathered around a magazine called *LM* (*Living Marxism*).

LM sought to cover up war crimes committed by Bosnian Serbs against Muslims, publishing an article by Thomas Deichmann claiming that the detention centres were a fabrication of the mainstream media, in this case the *Guardian* and ITN. (Deichmann was a witness for the defence at the trial of Serbian war criminal Duško Tadić.) The *Guardian* and ITN had told the truth. *LM* was sued for libel and lost the case and had to close down, but the cabal soon regrouped behind the Institute of Ideas (now the Academy of Ideas) and the *Spiked* network. *Spiked* is now funded by US oil billionaires the Koch brothers, who gave them $150,000 in 2016, the year of Brexit and the Trump election. *Spiked* promotes mass deregulation, including scrapping environmental rules and climate targets. Such rules, of course, impinge on the freedom of the present to slaughter the future.

One RCP member was Claire Fox. In the name of liberty, the RCP justified IRA murders, defending 'the right of the Irish people to take whatever measures necessary in their struggle for freedom' in a newsletter published shortly after the 1993 Warrington bombings which killed two children. Although Fox has said she no longer holds those views on the IRA, she has suggested that neither child pornography nor terrorism videos should be banned. 'It is those who have suffered the most who should be listened to the least' was one of *LM*'s lines. Fox has also been elected a Brexit Party MEP. Boris Johnson's government has just nominated her for a peerage.

Today's Libertarians support UKIP with its naked nationalism ('Believe in Britain'), and they cheer Trump's slogan 'Make America Great Again'. As the propaganda machine whirred, Martin Durkin produced *Brexit: The Movie* and made a fawning documentary about Nigel Farage. Steve Bannon says he began visiting the UK frequently in 2013, meeting Nigel Farage and James Delingpole and Raheem Kassam, an Alt-right nationalist who was senior adviser to Nigel Farage. Bannon hired Delingpole and Raheem Kassam to work for him. Kassam sought to repeal a ban on former members of the fascist National Front and British National Party from joining UKIP. He describes the Quran as 'fundamentally evil' and supports curbing Muslim immigration to the UK. Bannon was one of the covert forces behind Brexit, acknowledged by Farage on the day that Article 50 was triggered, when he thanked Bannon and *Breitbart*: 'You helped with this. Hugely.' Bannon was former Vice-President of Cambridge Analytica (CA): a whistleblower, Brittany Kaiser, says, at CA, 'I saw virulent racism and unchecked disinformation being channelled directly into voters' Facebook feeds.'

Racism is on the rise, at the heart of government. A former adviser to Boris Johnson, Andrew Sabisky (a protégé of Dominic Cummings), has argued that Black people have a lower IQ on average than White people, saying that the immigration system should be designed with attention paid to 'very real racial differences in intelligence'. Boris Johnson, who has made openly racist remarks himself, refused to condemn these views or to sack Sabisky. Johnson has chosen the Libertarian Munira Mirza to be head of the Downing Street policy unit and appointed her to lead the government's new commission on racial inequalities: she opposes multiculturalism, and has written on *Spiked* that institutional racism is 'a perception more than a reality'. Steve Bannon helped Boris Johnson draft his infamous column attacking Muslim women in *niqabs* as looking like 'letter-boxes' and 'bank robbers'. Johnson dishonestly denied his contact with Bannon, but video evidence demonstrates it.

'We will glorify . . . scorn for woman,' said the Italian Futurists, who called for the destruction of feminism. Today's Libertarians are only too happy to oblige. Andrew Sabisky has supported eugenicist views in arguing for enforced contraception at the onset of puberty for poor people. *Breitbart* reports an African cardinal comparing abortion with the Nazi Holocaust and yelps its support for Trump's relentless hatred of women. Calling breastfeeding 'disgusting', Trump was reported to have remarked of women: 'You have to treat 'em like shit.' So free is he of ethical conventions that he makes sexual comments about dating his own daughter, while Libertarian Mary Ruwart opposes restraints on child pornography. Regulations preventing child pornography, to Claire Fox, offend the right to free speech. At a meeting with young women, she told them that 'Rape wasn't necessarily the worst thing a woman could experience' and was then amazed that they were upset.

The Alt-right uses 'cuckservatives' as its hypermasculine insult to Republicans, while Alt-right sympathizer Daryush Valizadeh (Roosh V) emphasized brute physicality as a hypermasculine ideal and, in line with his gender-fascism, proposed to 'make rape legal if done on private property'. Marinetti, meanwhile, vilified the female body, sought a hypermasculine transcendence of humanity, to become 'the multiplied man'. For him, women are 'a symbol of the earth that we ought to abandon'.

'Would you rather your child had feminism or cancer?' asks Yiannopoulos: in 'Gamergate', he referred to 'sociopathic feminists'. The fact that women had received rape threats was to him not a serious issue, never mind that video-game developer Brianna Wu's home address was posted online: she received so many rape and death threats as a result that she had to move out of her home.

Those victimized by the Libertarians have often been silenced: hate speech will do that to a person.

Like Italian fascism's Il Duce, Libertarians support the 'Strong Men' – the rise of the Übermensch in the giddy upsurge of Trump and Putin, with Farage nibbling from their fingers and Le Pen in the wings. Rex Warner's airman finds a surrogate father, a kind of godfather, in the Air Vice-Marshal, and there is something of *The Godfather* in this – a touch of the mafia, which became so apparent in the Trump camp as he gave his family key jobs around him, while he treats women as baubles, never forgets a grudge and speaks a language of violence and revenge.

Tropes of Italian history are cropping up everywhere in the Libertarian story: Trump acts the emperor: Yiannopoulos arranged for his supporters to carry him shoulder-high on a 'throne' to address students like an emperor with his praetorian guard. There is a flavour of the decadence of the dying Roman empire – the flagrant, public, abusive sexuality which Trump boasts about, the wealth of empires at his fingertips. Yiannopoulos played for Trump the same role that D'Annunzio played for Mussolini. Yiannopoulos was the announcer only, the pretty messenger who would be tossed away when the main act came on stage.

D'Annunzio (1863–1938) was an attractive playboy, a poet of the Decadent movement, a racist, journalist and lover of the Übermensch – he portrayed himself as a

Nietzschean superman in film – and fascist sympathizer. He was a style icon of his times. A charismatic aesthete, he stood for election in 1897 as the Candidate for Beauty, and his aesthetics influenced the fascism of Mussolini, (including the use of the Roman salute), inflecting the machismo of the leader. Politics was performance, and he compared himself to Nero as artist-tyrant. D'Annunzio had an extravagant, audacious style, narcissistic and self-regarding. He was both a self-publicist and a liar: faking his death to ensure book sales. He was, like the Futurists, obsessed with aeronautics, speed, technology and war (his mansion has a battleship embedded in the side of a hill) and he was contemptuous towards what he called 'the stench of peace'. In his praise of the idea of flight, he wrote: 'Life on earth is a creeping, crawling business. It is in the air that one feels the glory of being a man and of conquering the elements.'

Like D'Annunzio, Johnson is a self-publicist and liar, a playboy journalist, scornful of women, adoring of 'Strong Men', narcissistic and decadent. I knew him once. Not well. We were at Oxford at the same time. What struck me about him was that there was no difference between playing and being: it was all a game. Nothing mattered. He went into politics with a willed ignorance of how most people experience the world. He didn't care. I see no difference now. The woman who recruited for the infamous Bullingdon Club during Johnson's time said that the members of the club 'found it amusing if people were intimidated or frightened by their behaviour'. They showed disdain and an air of superiority. The characteristics that Johnson displayed at Oxford, she said, which included 'entitlement, aggression, amorality, lack of concern for others – are still there, dressed up in a contrived, jovial image. It's a mask to sanitise some ugly features.' The decadence of the Bullingdon Club, its savage wasting of money, trashing places, its excess, has the flavour of a Roman orgy: fit for the emperors that these boys were bred to become. Little Neros.

D'Annunzio was obsessed with Nero. So was Yiannopoulos, who used the Twitter handle @NERO until Twitter banned him for life for abusing Leslie Jones. Both were flamboyantly image-obsessed. Yiannopoulos's talks are 'shows', his role as poster-boy for Trump, his decadence and his racism are all prefigured in D'Annunzio. Both have an element of the Trickster in them, the neither-nor character, skirting the edge of fascism, not quite in, not quite out. Trickster wants attention but not responsibility – to create chaos and then skedaddle, like Farage after the Brexit vote.

D'Annunzio was famous for ostentatiously sexualizing himself: he had a night-shirt made for him with a hole cut out for his penis, scarlet and gold tapestry around the hole, while his slippers were patterned with phalluses. This is a trope of proto-fascism: Putin, too, is photographed shirtless, striking his hypermasculine poses. Marine Le Pen uses a near-naked male underwear model alongside her for a photoshoot. For Yiannopoulous, artifice is his message, his meaning and his morality. He is obsessed with physical attractiveness – his own and others' – posing with photos of shirtless young men. His self-conscious aesthetic looks like fetish-fascism; he plays Wagner and reads works on racially stereotyped sexuality, while the rider for his college tour includes defuzzed peaches and roses without thorns: nature engineered to suit a jaded appetite. There is a decadent chiaroscuro in Yiannopoulos; the sleek, wealthy, well-tailored body versus the cruelty of his political views. It is echoed by Steve Bannon, who is openly fascinated by Mussolini because he 'was a guy's guy. He has all that virility . . . He also had amazing fashion sense, right, that whole thing with the uniforms.'

And when I say this isn't funny – when people in their millions do not think that fascism is funny – the Libertarian response is that we lack humour. As if it is all a game and the joker trumps all. The facts of fascism are not funny. And it is creeping right into the heart of government.

Today's Libertarians share with fascism a contempt for democracy and the things that support a democracy: an independent judiciary and a media that is free and honest. Trump threatens to seize a second term in office if he loses an election, and openly attempts to stop people voting by refusing sufficient funds to the US postal services. When all the US intelligence agencies are united in saying that Russia meddled in the US 2016 election, thus directly attacking American democracy, Trump sides with Putin, and calls the press the 'enemy of the people' (the term favoured by Lenin and Stalin.) In the UK, Boris Johnson prorogues Parliament: when the judiciary refuses to support Johnson, the *Daily Mail* called the judges 'enemies of the people'. In Brexit trade negotiations, a Tory minister openly admits that Johnson's government intends to break international law. Brexit was heavily influenced by Russia, and the Leave campaign broke electoral law. On both sides of the pond, the Libertarians in charge are flying in the face of legality, free of all shackles, including the restraint of law.

The Libertarian Alt-right fits the profile of the psychopathic mindset. It sees people as objects – units, if you like. Viewing speech without consequence, guiltlessly cruel, the psychopathic mind is dishonest, deceitful, power-hungry, lacking in empathy. (According to a 2011 study by Jonathan Haidt, Libertarians experience less empathy than others, and are less disturbed by violence.)

A crucial way in which the human mind practises empathy is in the use of metaphor; not as a literary device but as a stance of psyche, a way of understanding otherness and making stories of connection and relation. Metaphor is weirdly missing in the Libertarian screeds: the literalism is choking. (Oddly, one metaphoric use of language is the actual title *Spiked*, with connotations of date-rape drugs and also of journalism too poor to merit publication.) Mostly, though, metaphor is missing. Metaphor comprehends the world of spirit and imagination. It deals in pluralities and the unmeasurable. It doesn't seek to deceive but rather to augment truth. It is essential to all kinds of understanding. Metaphor is the opposite of rhetoric and works like anti-fascism in the human mind.

If Libertarianism has an ideological role model, it is Julius Evola, anti-semite and godfather of Italian fascism, whose ideal order was based on hierarchy and race. Richard Spencer, the Alt-right White supremacist who led the 'Hail Trump' fascistic posturing at a Trump election celebration, invoked Evola's vision of the Solar Civilization, a reawakening of Whites, the 'Children of the Sun'. *Breitbart* cites Evola as key to the rise of the Alt-right and Steve Bannon referred to Evola as an important influence.

If Libertarianism has a mythic model, it is Deus Invictus, the 'god unbound', who loathes any tether, shackle, or constraint. Invictus was the epithet applied to the supreme deity Jupiter, Übergott, and to Mars, god of war, and to the empire-building Caesar. It is associated with the triumph of individualism and totalitarianism, as well as solar monism, a sky-god, singular as the sun, and unbound from – and hostile to – the pluralities of the land. Deus Invictus drove the Italian Futurists' demand to be free of the 'yoke' of the past and loosed from ties to the natural world.

Deus Invictus lies behind the Libertarian refusal to be shackled by cuts in carbon emissions, as behind their demand to be free to fly, and their obsession with anti-natural technology. Deus Invictus is god of the Libertarian-supported Singularity. The Deus Invictus complex drives unfettered capitalism and the unregulated industry that allows some to live like gods by making others live like cattle. Characteristics such as kindness, modesty, slowness, generosity, receptivity, pity, honesty and truth are detested by Deus Invictus, for they are the fettered emotions. In honesty, you are bound to tell the truth. You are tied by respect, linked to others by love, tethered by kindness to kinship with nature, and restrained by a sense of justice and conscience. Libertarians loathe political correctness because it puts the brakes on bigotry, restricts racism, reins in sexism.

In the dying days of the Trump regime, he infamously roused a mob to fury by repeating his lies that the election was stolen from him. Being unconstrained by truth was Trump's most powerful weapon, as fascism loves lies as much as violence. At that rally on 6 January 2021, his incendiary language, invoking the Strong Men trope, was aided by his attorney, Rudy Giuliani, calling for 'trial by combat'. Trump incited his thugs to go with him to storm the Capitol. He himself left them, going to watch it on TV. With the window-smashing, the fatal gunshots and the smoke bombs, the poisonous fumes of fascism were not so much smelt seeping under the door as seen swirling right within the citadel of American democracy.

Like Trump, Mussolini incited his Blackshirts to march on the centre of government in Rome; like Trump, he did not join them but went off safely elsewhere. The event instigated Mussolini's dictatorship. Trump attempted the same, ensuring a bloodied pseudo-coup would instigate his dictatorship of an alternative – Alt-right – universe in which he would be god, and where his fascist legacy would outlast him.

In the decadent days of the late Roman empire, Deus Invictus, as patron of soldiers, was shown with a whip and a globe to emphasize dominance and invincibility; his shield was spiked. Typified in Libertarianism and personified in Trump's solar solipsism, with his backdrop of gold curtains, Twitter-roaring against the unbearable restraints of truth, respect or social justice, Deus Invictus is a ruthless enemy to the living world, the god unchained to scorch the earth. What has been set in motion is an ideology of monoism without plurality or otherness, furious for its own freedom, an idiot divinity unleashed upon the world.

The Forests of the Mind

Eros is coursing through the forest. The forest is mewing with its jaguar life. Life is spiralling into poetry.

I am in the other world, I thought, at once in the actual forests and in the forests of the mind, where the visible world is not denied but augmented.

I had gone to the Peruvian Amazon seeking treatment from forest doctors for an episode of depression so long and so severe that I had worked out how I was going to kill myself. (Wrists. Length-wise. In the bath.)

What I experienced was more than the healing of this desolate madness, it was a sense of the raw, green-eyed, lustrous sacredness of life which has never left me, and which came through a sense of identification with other creatures, the knowers of the forest. Shamanism is universally concerned with the well-being of both nature and human nature and the relationship between them, and here in the Amazon, the shamans' rituals unleashed in me a force of empathy exact, sensitive and enraged. The pull of my imagination was tautened by all the aspects of the occasion – the medicine, the night, and the shamans' songs – until, the torque twisted most surely, I lost my singular self and stepped across the border in a wild, charged, ferocious apprenticeship to a jaguar.

At times, I felt a hot sexuality coursing through me, as if, in pelt and paw and breath, I could feel from within my body a radical love for the Earth as strong as the gravitational force. At others, I felt a prowling rage as a jaguar might feel, watching the trenchant stupidity of deforestation. My society is destroying this forest. The anger burnt me and I could feel the fire of pure fury. How can modernity know so little for knowing so much? The sacral need to protect life in all its forms swept through me like wildfire, with meaning not only intellectual but physical, sensory, what Hermann Hesse called 'this felt faith'.

This was shapeshifting. It is part of the repertoire of the human mind, cousin to mimesis, empathy and Keats's 'negative capability', known to poets and healers since the beginning of time, the beginning of mime. It did not have literal truth, quite obviously, but had a 'slanted, metaphoric truth' – the words I used, when the page was printed, to describe it.

Shapeshifting is a transgressive experience, a crossing over: something flickers inside the psyche, a restless flame in a gust of wind, endlessly transformative. The mind moves from its literal pathways to its metaphoric flights. Art is made like this, from a volatile bewitchment of a self-forgetting and an identification with something beyond. Out of this is born a conviviality with everything alive, the relationship acknowledged and the necessity of its protection vouchsafed.

Ted Hughes once said that the secret of writing poetry is to 'imagine what you are writing about. See it and live it . . . Just look at it, touch it, smell it, listen to it, turn your self into it'. If I had met him, I would have wanted to kneel, an ancient fealty due. One writing exercise Hughes suggested for students was titled: 'I am the Amazon'. We are what we think, and we humans have a way to become other, in a necessary, wild and radical empathy.

Shapeshifting involves a willingness to make mimes in the mind, copying something else. Art, meanwhile, depends on mimesis furthering our desire to know and to understand. In a recent, Ovidian, dance piece, *Swan*, French dancers performed and swam with live swans, imitating the birds in a mime which alluded to the metamorphosis of all art, and to the artist's ability to lose themselves in order to mirror this something beyond.

> But we, when moved by deep feeling, evaporate; we breathe ourselves out and away;

wrote Rilke in his Second Elegy.

In making art, the artist expires, breathing themselves out to allow the inspiring to happen, the breathing in of the glinting universal air, intelligent with many minds, electric and on the loose. Artist, shapeshifter, shaman or poet, all are lovers of metamorphosis, all are minded to vision, insight and dream.

Self-appointed shamanism can reek of cultural appropriation, but even in cultures which have temporarily misplaced their shamanism, the role survives, donning a deep disguise. Joseph Campbell and others believed that artists have taken up the role, and it seems to me that this is true for a particular reason, that both art and shamanism use the realm of metaphor, where emotion is expressed and where healing happens. With the metaphoric vision, empathy flows, knowing no borders. Both artist and shaman create harmony within an individual and between the individual and the wider environment, a way of thinking essential for life; poetry works 'to renew life, renew the poet's own life, and, by implication, renew the life of the people', wrote Ted Hughes. But ours is an age of lethal literalism which viciously attacks metaphoric insight and all its values, an age which burns the Amazon and mocks those who would protect it, sing it and become it, an age which physically destroys the sacred sites and art of Indigenous cultures. A cave, sacred to Indigenous Australians in the Juukan Gorge in the Pilbara region, that demonstrated 46,000 years of continual occupation through the last Ice Age and into living memory, was exploded in seconds by the mining company Rio Tinto. They simply detonated one of the oldest sacred monuments in the world. Blew it up in a few minutes, one day in 2020.

Millennia ago, an unknown artist painstakingly chiselled a human face out of rock, carving huge, haunting eyes and an expression vivid in intensity. It is probably the oldest artistic representation of the human face anywhere in the world. It is one of hundreds of thousands of rock carvings spread across eighty-eight square kilometres in Western Australia's Burrup Peninsula, known as Murujuga to Indigenous Australians. Some carvings are at least 30,000 years old, and the site may be twice the age of the famous, enigmatic cave paintings of Lascaux, France, which are perhaps 20,000 years old.

The carvings jump with life – outstretched hand- and footprints, birds, wallabies, emus, whales, turtles. They refer to stories that reach back beyond the Ice Age, depicting creatures long-extinct, such as the fat-tailed kangaroo and the Tasmanian tiger; to Indigenous Australians, this is a 'prehistoric university'. The engravings speak of the Dreaming, that subtle, diffuse, living past, re-created in the present through song. This rock art is implicit shamanism, evoking spirit and sacred energy, metaphorical and metaphysical. The area is being destroyed by a literal energy company: Woodside Energy Limited.

Shamans may be feared if they are seen to 'bind' people with 'spells', and a good storyteller is spellbinding. People are entranced by Debussy, mesmerized as by a magician. The shaman may sometimes act the part of a showman and, from Liszt to The Beatles, performers are glamorous, in a history older than they may know, for 'glamour' was an old term for 'bewitchment'.

Artists – for which read musicians, dancers, performers, sculptors, painters, directors, writers and poets, the lot – often suffer an overwhelming psychological experience in youth, commented Joseph Campbell, when it is as if 'the whole unconscious has opened up and they've fallen into it'. Shamanism, like art, is a calling, and a young person may be 'doomed to inspiration', as anthropologist Waldemar Bogoras wrote of the Siberian shamanic vocation. In a painful transformation lasting months or years, the young shaman loses interest in life, eats little, is withdrawn or mute, sleeping most of the time. It reads like a portrait of the young artist in a devastating depression. The young shaman overcomes the illness through the practice of shamanism, just as many artists know that their own best medicine is found in their work.

Art creates an emotional catharsis, said Aristotle, which rebalances our emotions. With the shamans in the Amazon, I felt a powerfully cathartic effect, as they used a medicine called (amongst other things) *La Purga*, the purge, which is one translation of the Greek word 'catharsis'.

The shamanic trance is like the entrancement of the artist, when the ordinary laws of time are repealed and the illumination of daylight doesn't apply. As intense as the flame absorbed by its own burning, as the wine intoxicated by its own alcohol, as the wind swept by its own gust, the shaman's paradoxical role of ferocious power is coupled with unshieldable vulnerability.

Shamans and artists alike occupy an ambivalent place in society. Shamans can be hated for their power, as artists can, treated to acts of savage psychological violence, and yet are also treated to acts of psychological reverence given to no other profession.

Shamans have traditionally lived on the edge of their communities, and the quality of 'edge' is what marks original artists. The shamanic path and the artist's way are both associated with the hero's journey, as Joseph Campbell terms it, 'the dangerous, solitary transit'. Solitary. That's the word. The path known to be stony and lonely, the unknown destination ever beyond. In what is understood to be a self-portrait, Michelangelo painted the solitary, sad figure of a centurion in *The Crucifixion of St Peter*, and the young William Blake recast the figure, deepening the loneliness to accord with his own experience of being a visionary in a scornful world. 'Wisdom is sold in the desolate market where none come to buy,' Blake wrote in *Vala or The Four Zoas*, in a bleak version of a universal understanding: as Inuit shaman Igjugârjuk in the early twentieth century commented, 'True wisdom is only to be found far away from people, out in the great solitude.'

During one ceremony in the Amazon, I had the sensation that one of the shamans had, as it were, sent his soul out to find mine. Although I was lost in the dark forest of depression, suddenly he was there, in a bright, clear pool, healing and sunlit. Shamans use the term 'soul-loss', which was not an expression I had heard before, but this was exactly what I felt the moment mine was found. A good healer of any kind can find people who are lost in the forests of the mind.

Halfway through his journey in life and lost in a dark forest, Dante begins his poem-path. By naming his lostness to his readers, they, if they are lost themselves, may feel understood – found – by him. Artists send their soul out into the world in a parabola, thrown from the heart of solitude so that in the arc of its return it can comprehend and speak to the loneliness and separateness of other minds. A book, as Franz Kafka said, must be an ice axe to break the sea frozen inside us.

With his raw materials of rough magic, wax, felt, fat and coyote, Joseph Beuys interwove art and the shamanic role. Max Ernst took on a shamanistic familiar, the bird-king Loplop. Novalis wrote of Romantic poetry's aim to 'give mysteriousness to the common, give the dignity of the unknown to the obvious, and a trace of infinity to the temporal'. He could have been describing the shaman's art.

Yehudi Menuhin came to consider his playing a form of healing. Sámi shaman-musician Nils-Aslak Valkeapää and singer Woody Guthrie (both political artists) shared a parallel experience when they were young: people began coming to them for help and for healing of the psyche. They responded instinctively and empathetically, taking the role of healer, as they would in later life heal both the individual and the body politic through their music.

Shelley considered poets the unacknowledged legislators of the world. Shamans may work as acknowledged legislators of their worlds, and part of the shaman's traditional role is to regulate hunting, laying down rules, taboos and off-limit places or times, in order to allow wildlife to thrive. Incidentally, Welsh bards were actual legislators as well as poets, and in fifth-century BCE Athens, theatre and music were tied to governance, so attending performances was considered part of a citizen's preparation for jury service and legislation.

Anthropologist Gerardo Reichel-Dolmatoff (in the fascinating anthology *Shamans Through Time*) describes Amazonian shamans as typically curious, humanistic and fascinated by myth and tribal tradition. A shaman's spirit will 'illuminate', it should 'shine with a strong inner light rendering visible all that is in darkness, all that is hidden from ordinary knowledge and reasoning'. Lit by their own sun, like Van Gogh, artists are guided by their own vision from the dark side of the mind.

Mediating between a world of daylight sight and a world of night insight is the role of both shaman and artist. Rilke termed it 'divine inseeing'. Offering a particular kind of attention yields a different kind of knowledge: in part it is the wisdom of the dream. Amazonian shamans may be called *sueñadores* or *sueñadoras*, the dreamers. 'I am here to dream dreams,' said Nils-Aslak Valkeapää.

Hermann Hesse wrote: 'I stand alone in my role as a "dreamer".' At night, when the day closes its eyes, othersight is possible. With our eyes closed, we see in dreamsight, sharing nightly the paradox of vision known to Tiresias, the blind seer. It is revealing that the first recorded metaphor is in the oldest written story, *The Epic of Gilgamesh* (2700 BCE), where sleep steals up 'like soft mist'. From the earliest literature, metaphor relates to dreamers.

Wallace Stevens calls poetry 'The Necessary Angel', and the root of 'angel' is the Greek word for 'messenger'. To be a messenger, to negotiate between the real and visible world and the true and invisible world is, shamans say, a crucial part of their role. The artist, too, is a messenger between actuality and imagination. Hermann Hesse said the poet is a messenger 'between the familiar and the unknown as if he were at home in both places'. The kinetic power of artist and shaman resides in their ability to return their insights to their communities, to go down with Dante and find the providential paradox that only in the depths can the highest ascent begin.

It is a peregrine part, shamanism and art both. Sometimes the journey is juddered, as the Man from Porlock cannons through the door. Sometimes the journey is a seduction and sometimes a refuge, often more real to the artist than any other place. For a long time 'Thou must pass for a fool,' writes Emerson. 'And this is the reward: that the ideal shall be real to thee.' The writer-character in Hemingway's *The Garden of Eden* begins to live more and more in 'the other country', the land of his novel, and the real world comes to seem false. Dwellers on metaphor dwell more truly in that other world.

If I were asked what is the greatest human gift, I would say it is metaphor. A little boat of metaphor chugs across the seas, carrying a cargo of meaning across the oceans that divide us. Metaphor is how we relate to each other and how our one species attempts to comprehend others. With this gift, humans listen and speak more intensely and the meanings of all things – ocean or forest, snail or chaffinch – grow outwards in concentric rings of concentrated word-poems. 'Every word was once a poem,' said Emerson, and 'language is fossil poetry.' So a tulip, for example, ultimately derives from the Turkish word for 'turban'.

Metaphor works with the legerdemain of the psyche, the lightest of touches to shift the mindscape, transforming one thing into another, leading to new ways of seeing. Metaphor follows Emily Dickinson's injunction to 'tell the truth but tell it slant', so, slantwise by Saturn-mind running rings around literalism, metaphor is a canted incantation, it breathes life into fact, it enchants. And metaphor is the language of the shaman and the artist.

Shamanism and poetry both use indirect, implicit and enigmatic language to know and to heal. Amazonian shamans, talking to anthropologist Graham Townsley, describe their expression as 'Language twisting-twisting', explaining its elliptical and abstruse power thus: 'I want to see – singing, I carefully examine things – twisted language brings me close but not too close – with normal words I would crash into things – with twisted ones I circle around them – I can see them clearly.' Metaphor, says Townsley, changes the world by changing people's perceptions of it. Artists prefer to work in subtle, oblique ways, knowing that blunt and obvious references militate against deeper thinking.

At my most depressed, in order to describe what I felt, I said I was drowning. No literal description was good enough. Only a metaphor seemed a craft strong enough for me to cling to, a boat to carry my grief across the sea to someone else's mind. For some, elliptical language may be the only way to unwrap tight, compressed pain. For others, self-disclosure can take place only in language which conceals, even as it draws attention to itself. A good therapist listens carefully to a client's metaphors, ellipses and masks, for they are true beyond literalism.

Halfway through his path in life Jung found himself lost in an underworld of the psyche which he was determined to understand. He noted two influences: one the 'spirit of this time', concerned with 'use and value', while the other 'rules the depths . . . the inexplicable and the paradoxical . . . the melting together of sense and nonsense, which produces the supreme meaning.' Shamanic thinking in the metaphor world.

The shamans I visited used a metaphor common in the Amazon: you have been struck by arrows, they said, poisonous darts designed to kill the spirit. It was a perfect metaphor for what I – like so many artists – had experienced. And, they said, they could suck them out of my mind, so, like powerful dramaturges, they dramatized the metaphor, embodied its meaning, staging the powerful sense of cure, sucking the poison out of my head. It made me well.

Placebo effect, a cynic may say. Absolutely. The word has its roots in 'pleasing', and good medicine, like good art, should please in order to heal: the placebo's success is evidence for the power of metaphoric medicine to heal mind. 'My project,' says the shamanic Prospero in *The Tempest*, is 'to please'. The shamans sang songs over me called *icaros*, half-whistled, half-voiced, half-heard, half-imagined: exquisite Ariel music, in themselves mind-medicine, curative music sung by these *curanderos*, as they may also be called. Ted Hughes writes of 'the healing effects of reading and writing poetry', which recalls the fact that the ancient Greek god Apollo was god of poetry, music and healing.

A bundle of eagle feathers perhaps, or a pelt of wolf skin, a drum painted with ciphers of a particular land, a hare paw, maybe, or a tiger tooth or bone of reindeer, shamans identify themselves with the motley Earth, in patchwork cloaks jingling with diverse life. The striking correspondence of shamanic practice the world over, the similarity of costume and the likeness of role have led some to suggest that shamanism somehow 'began' in one place and spread globally. To me, that is absurd. Rather, it will arise, this ur-religion, wherever earth meets mind.

Wordsworth called himself 'nature's priest', and the term suits a shaman's role from the Amazon to the Arctic. In Tanzania, shamans may be called 'doctors of the forest', and they must protect it. Shamans have always known what the discipline of ecology has painstakingly re-taught: that everything existing is interdependent and coursing with a transcendent and sacred life force. 'So everything is necessary. Every least thing. This is the hard lesson. Nothing can be dispensed with. Nothing despised,' writes Cormac McCarthy, in *The Crossing*: 'We have no way to know what could be taken away. What omitted. We have no way to tell what might stand and what might fall.'

Shamanism survived in Britain and much of Europe until the witch-hunts, when the beyonders' wisdom and nature-knowledge was hunted down. So knowers and seers had to shift their shape, in a new transformation which radically masked who they truly were.

I think I can see it happen. Watch Shakespeare.

His fools and jesters shine out of his plays with the illumination so typical of the shaman role. They are charismatic, mercurial, they play as reckless tumblers and clowns and yet so often demonstrate their primary paradox: that the Fool is wise. The quality of empathy which shaman and artist both know is portrayed in Lear's poignant, devoted, suffering fool, and is played out in numerous acts of mime and mimicry, those related skills of fools. Characters of licence, they can speak truth to power, they are called mad when they speak their sanities to a crazed court; they may, like shamans, speak a language twisting, twisting the words of a king to wring out his honesties.

Shakespeare's fools trip logic with inspired nonsense and, acting by their own twilights, can express truths in a shadow language denied to the glaring light of literal speakers. To restore balance, they create a turvy-topsy world, they are grave at comedy and witty at the graveside, they are liminal, living on the edge. Like shamans, they are unsalaried and work by pleasing, by placebo, even when their healing stings.

Lear's Fool or Feste or Touchstone come alive in the between-spaces, the manoeuvrings of love, of power, of psyche. Shakespeare's fools often speak directly to the audience, playing on the stage between the Play and the Real World (how I hate that half-witted term which denies half the mind its metaphorical realities). They perform their motley antics right on the edge where metaphor plays, the edgy dimension of the player.

Shakespeare is profoundly shamanic, a magician who knew how magic was related, etymologically and intrinsically, to imagination: he was that great magician obscured in the forest which Orlando speaks of, living in necessary concealment. Some say that Prospero is modelled on John Dee, alchemist and seer, while others say he represents Shakespeare himself. He is, to me, a portrait of shapeshifting made just when shapeshifting itself cast off its old costume, just when the shamanic role was far too dangerous and must, like Prospero's cloak, be cast away. For the tide had turned and magic was a castaway, the magician shipwrecked on the dry island of Protestant literalism. What was needed was a sea change, and as Prospero abjures his 'rough magic' of mere matter, actual materiality, in the literal world, he stands before us a *metaphorista*, on a different shore, with a poet's power over the mind's ocean, the imaginal world.

Listen: the actors are changing behind the scenes. At the threshold you can hear their shifts rustling. And just at that moment, on stage, Prospero slips off his cloak, and Shakespeare shifts the shape of shamanism into art, the magician becoming the imaginer. A threshold character at a threshold moment in history, Shakespeare's genius gathered the first and most august harvest.

Just in the nick of time.

For the curs of Puritans were there, snarling at the gods.

They closed the play-houses of theatre. Seeing everything in black and white, literal as black letters on the white page, they wanted to close the play-houses of the mind, too, where imagination makes a play on words and thoughts, making ludic allusion to an illusory world. Malvolio, ill-wisher to revelry and liveliness, bleaches the motley fool of his colour. No enthusiasm. In its fossil poetry, enthusiasm, *en-theos*, means the god within, and that unmediated – natural – divinity was cursed. No metaphorical revelations, only literal scripted teachings. And no icons. The alchemy of art, which transmutes the base metal of literalism into the gold of metaphor, was denied. Literalism was in the ascendance. Catholicism's metaphorical view was erased by Protestant literalism, which ran its writ over the following centuries through the ethos of the Industrial Revolution and Utilitarianism. The human psyche, understood so well by artists and shamans, was reduced to a machine by dire literalism to the point where Descartes located the psyche in the pineal gland.

Such mechanistic thinking continued its brutalization as if mind didn't matter, and in the 1930s depression was treated by leucotomy, a form of psychosurgery where the frontal lobes would be severed from the rest of the brain, as a cure. It was done, says medical historian Roy Porter, 'often using an ordinary cocktail-cabinet ice-pick, inserted, via the eye-socket, with a few taps from a carpenter's hammer'. Instead of the healing power of metaphors (literature an ice axe to crack the frozen seas within us) a literal ice pick was shoved through the literal eye into the psyche.

In the Amazon, I met healers who would not call themselves shamans because, barely fifteen years before I was there, literalist Christian missionaries had urged Indigenous people to kill their own shamans for their othersight. Twentieth-century Huichol shamans in Mexico were murdered at the behest of that same faction of the church. Historically, Siberian shamans were interned and executed by Soviet authorities, their costumes and drums burned. Those who said they could fly when their minds were metaphorically winged found that they were persecuted with literalism's sadistic mockery, and flung out of helicopters. In northern Scandinavia, Sámi shamanism was subject to long episodes of persecution for hundreds of years and was driven underground as the spiritual and secular authorities sentenced shamans to death. Bolsonaro today aggressively supports evangelical missionaries in their assaults on Indigenous land, bringing physical illness, including Covid-19, and a disease of the mind: literalism. The shamanic worldview is a key target for literalist Christians who would kill the very core of metaphor at the heart of the forests, where metaphorical thinking thrives in the tree-shade, earth-dreams and night-thoughts of forest life. Deforestation itself destroys the mind-environment that nourishes metaphor.

Rationalism, reason alone, is insufficient to the wholeness of the psyche, said the Romantics, and Blake contrasted the 'Man of Imagination' with the 'Idiot

Reasoner'. The role of shaman was safer described as 'art', but not safe enough: the Romantics were scorned for their shamanic willingness to translate the voices of nature by deaf literalists, the idiot reasoners who endumb themselves numb.

The Romantics represented the ascendancy of the brain's right hemisphere: an instinctive, metaphorical view which comprehends the whole, honours life, art, humour and metaphor. But modern Western culture privileges the left-hemisphere stance, which is an important servant (good at logic, language and engineering) but a terrible master. Seeing the timber but ignoring the forest, it counts the profit but discounts its own destructiveness. It has no regard for thoughtways other than its own. Short-sighted and narrow-minded, it cannot see the stupidity of its own position.

Fundamentalist Christians attempting to translate the Bible into one Amazonian language got stuck on the 'good Shepherd'. There aren't any sheep in the Amazon, so what were they to do? Unable to comprehend metaphor, they imported a few surprised sheep into the grassless Amazon, to the bewilderment of the sheep, people and forest.

Max Weber characterized Western modernity as a 'progressive disenchantment of the world'. This is an age of fundamentalist literalism in culture, as much as in religion or politics. Literalism, a dogma treasureless and lamentless, acknowledges only the coins which can be counted, frankly pounding the dolour of marked money. As Jung remarked, the spirit of this age understands use and value. Nothing else counts. Neither nature nor poetry. Barren postmodernists scorn the fertile Romantics. The cursed evangelicals kill the curers of the forests. The hate-filled Libertarians mock environmentalists for their shameless willingness to love the Earth. You catch the glare: a blinding sheen where shadow should be seen; the cold touch of steel where the warmth of wood should be felt; one dimension of scale where seven kinds of music should be heard.

Where is Basil the Blessed when you need him? One of the great holy fools of medieval Russia, Basil robbed the rich and gave to the poor and Ivan the Terrible built a cathedral in his honour, with jester's hat cupolas. In shamanism's wise and witty fooling, grave laughter mocks injustice. Mexican government troops were 'bombed' with hundreds of paper aeroplanes, sent by the Zapatistas, whose spokesperson, for the protection of Indigenous cultures and the forests, is the masked, wise, ludic, elusive, serious jester Subcomandante Marcos. Clowns Without Borders, beginning in Barcelona, has now mushroomed internationally. The Italian satirist Dario Fo, who used commedia dell'arte and mocked those in power, is now joined by Grillo, the comedian subverting politics. In Britain, carnivalesque protesters set up the Clandestine Insurgent Rebel Clown Army, and Rupert Murdoch had a custard pie thrown at him by a clown-protester. In Iceland, the comedian Jón Gnarr and the Best Party have taken mayoral power. In the States, the Yes Men mimic public folly, impersonating loathsome entities to tell the truths of corporate lies, posing, for example, as representatives of Dow Chemical to issue an 'apology' for Bhopal. Extinction Rebellion New York chose April Fool's Day to impersonate Google, offering a spoof announcement that Google would stop funding climate denial.

All clowns have the immunity of the court jester; the suits cannot hold them still, they are slippery as a bladder on a stick. 'Who shall bring redemption but the jesters?' asks the Talmud.

Jesus-the-jester subverted traditional authority; St Francis, 'God's Jester', was a holy fool, his happiness simple and silly, in fact, as that interesting word has it, which derives from *saelig*: holy. For calling Earth Gaia, a goddess, James Lovelock was called a holy fool – an insult to treasure.

A woman walks into a bank.
– Can I change some money?
– What do you want to change it for?
– Dreams.

We are such stuff as dreams are made on, said Prospero, while in a disenchanted modernity a delinquent prosperity is loosed upon the world where only materialism matters, and that only for the few. Utilitarianism, efficiency and the profit-motive are literalism's tyrannies, a totalitarian state of mind deplored by poets, including Paul Kingsnorth, who, in a furious refusal of The Useful, writes: 'I have hid my heart in a butterfly.' Eros tucked inside Psyche, for the butterfly was the Greek's image for Psyche.

For the Greeks, Eros is a liberating force, unleashing the Psyche's creativity in art. In Greek myth, Eros cannot be separated from Psyche. Dante considered that the interplay of stasis and movement was the result of the gravitational attraction of love which 'moves the sun and the other stars'. The human psyche cannot be unbound from the force of erotic gravity, a jaguar's ferocious love for the felt earth, a love which prowls in the human heart no less, a helical love spiralling inwards to this erotic earth, a love which transcends downwards, earth-enchanted and grave with fury at the burning of the forests.

There are shooting stars in daylight, as well as by night, and the ones I like the best are the ones you never see, the stars which play truant from ordinary sight and can be seen only by psychelight as they fall upwards from below, towards the Earth, rising in the erotic gravity of love.

Happiness, Animals and the Honeyguide Rule

One morning in spring, when I was about twelve, at the edge of adolescence, I woke up with a heavy heart and a dilemma. It was a Saturday morning with no plans. I was aware of the immense peer pressure which dictated that I should be doing girly things like making clumsy forays into eyeshadow or trying on stupid clothes in front of mirrors. I resisted that pressure with all my heart. I didn't have a mirror or eyeshadow and the only clothes I wore, beyond my school uniform, were a pair of jeans, one brown T-shirt and one brown sweatshirt. The very idea of bras made me yelp with horror like Bart Simpson in the lingerie department. I was a total failure as a girl.

That day, I realized that to grow up female seemed to mean eschewing the natural world, while to be a child was to be a boy, to grab a fish net and a jam jar, fetch my bike and head for the ponds. That is what I did but, although it was springtime and I was in the same season of my own, I felt sadder than I ever had in my life, for it was valedictory, it was autumn in April, it was twilight at dawn. I sat crying, until I grew fascinated and consoled by a water boatman. His name was a metaphor, carrying meaning across from the human world to that of the insect. Each tiny leg of his made a little hollow in the water as he rowed his oars across the whole pond-ocean with a meniscus for a boat.

I was comforted by an insect. A water boatman showed me the way. Such is the reach of biophilia, of the necessity of everything, no matter how tiny.

Nature-connectedness is correlated with emotional and psychological well-being. From the Japanese 'forest-bathing' to the joy inherent in Norwegian *Friluftsliv* or the rush of oxytocin in dog-owners, gazing into their dog's eyes. Happily, dogs also feel the same oxytocin's delight when they are with the human they love. It is easier, of course, to love one's cat than to care about the chestnut clearwing moth or the rufous-fronted laughing thrush (subspecies *slamatensis*), but pets themselves can be the ambassadors of the natural world, leading us by the paw into a world richer and more minded than we could ever know by ourselves.

In the Sumatran rainforest, the long-tailed macaques leap like embodied jokes, making the very trees laugh with their sense of swing. A baby monkey jumps from liana to liana, curls its fingers around a branch and dives into a stream: aerial then aquatic acrobatics. A gecko runs up a buttress flank of mahogany and freezes, alert, silently glued to the trunk, its tiny tongue licking up termites. High in the trees, a Thomas leaf monkey, with its long white tail, whiskers and a Mohican, blinks and gazes, blinks and gazes.

This forest, filled with the jungle music of crickets and frogs, is home to all the creatures of *The Jungle Book*. I'd been invited to join an ecotourist trek to see orangutans, a critically endangered species. The hope of seeing one was only a part of my delight: to put it simply, forests make me happy.

When a wild landscape is lit with birds and ribboned with animal presence it tells us that all manner of living things are well, and it draws us inextricably into a shared happiness, whether in a savannah or rainforest or the woodland humming with joy evoked by Tennyson's lines of 'doves in immemorial elms / And murmuring of innumerable bees'. Thus the giraffes who caress each other with low hums, a gentle evening song of the envoiced world. Thus puffins, clowns of the air, possibly the most visually cheering of all birds. Thus rats who, if tickled, chirp like children laughing, while bonobos, if tickled, laugh until they fart. Laughter is a signal, a form of communication that tells others that the laugher is not only happy but wishes to spend more time with the laughee, welcoming the interaction as reciprocal exchange. When we respect the fact that all species are necessary to the well-being of an ecosystem, this sense of shared happiness can potentially include everything, from the spotted hyenas of the African savannah to the leeches dropping down on to us from the trees as we walk through the forests of Sumatra, or the huge spiders that return to my house every September for shelter from the first chills of autumn. Everything. Even the ticks and the wasps.

Orang-utan means 'person of the forest' from Malay and Indonesian. These great apes are highly intelligent tool-users, making a bed for themselves every night, using leaves to amplify their 'kiss squeak' sound of annoyance, while some orangutan individuals have been taught sign language. Orangutans are the first animal species (apart from humans) known to use 'calculated reciprocity', keeping track of gifts given and received, and weighing up the costs and benefits of gift exchanges. Though orangutans usually travel through the forests alone, and lead more solitary lives than most great apes, the mother–child bond is very strong. The sight of a baby orang-utan, a little person of the forest, cannot but make a human happy. I dearly wanted to see one.

In the Gunung Leuser National Park, a World Heritage Site, I was assured I was part of a reciprocal exchange. Ecotourism here, say both UNESCO and local people, is a way to protect these forests for the orangutan, to provide livelihood for locals and so lessen the drive towards deforestation for palm oil plantations. It was also, so they say, a vital way to take care of abused and mistreated orangutans, the most terrible case of which is a female orangutan who had been captured, kept in a brothel in Kalimantan, shaved and chained for sexual purposes. The National Parks sought to protect both animals and the forests themselves. I was guided by faith, hope and happiness.

I have never been so misguided. The guides constantly called out to the orangutans, who don't like being disturbed. Stressed by the shouts, they start to leave the area, and in their movements the guides can catch sight of them. 'We do it *because* they don't like it,' one of the guides said candidly. The guides want to make the tourists 'happy' by making the orangutans unhappy: it is not a good exchange.

Breaking one park regulation after another, the guides gave food to the orangutans and then brought a large group of tourists not only far too close to a mother orangutan and her child, but between them. One of the guides then began provoking the mother, leaping at her, goading her and laughing. Another stood over her with a stick, waving it in her face. She was distressed. I was angry and walked away, telling the other tourists that none of us should be there. I've never knowingly been complicit in an act of gang cruelty and I was horribly upset.

It was only when I got home, though, that I found out about worse aspects of tourism in Bukit Lawang, the main town that tourists stay in before beginning a trek of a few hours or (in our case) two days. Cannibalism is incredibly rare among orangutans, or indeed any great apes, but there are two documented cases of cannibalism among the orangutans, both in the vicinity of Bukit Lawang. While it is not possible to state the reason with certainty, the most probable cause, say experts, is the stress of uncontrolled tourism. Both cases of cannibalism took the form of mothers eating their infants.

When I learned this, far too late, I felt sick. No journey I have ever undertaken has left me so ashamed or so unhappy. It broke the honeyguide rule.

There is a bird called the honeyguide because it knows where the bees' nest is. But it cannot actually get the honey out on its own. It needs human hands to break open the hive, while humans need the bird to guide them to the right tree. Hence the honeyguide rule: that a happy human–animal relationship depends on reciprocal exchange. The honeyguide is an emblem of the best kind of relationship between humans and the more-than-human world, leading the psyche to sweetness.

When the relationship is at its best, creatures can draw us humans into their worlds, from the water boatman to a pet cat or a wild deer. They are the filaments of our tenderness, the reach beyond our grasp, the extension of our empathy, the wings of our minds. To be happy, the senses need to be stimulated and, through animals, our senses grow: we can extend ourselves out through their senses into sensory worlds of unquenchable richness. Each creature lives paws-outwards, reaching into the world with arms, tendrils, tentacles, antennae and fingers. Compared to many animals, we are beginners in the sense of smell, newcomers to sight, inexpert in the audible realm. For our own safety, we need to listen to their sensed communications as, before a tsunami, for instance, they listen to each other and, heeding warnings, can save themselves. There is a different kind of tsunami coming, but theirs are still the voices (or the silences) we need to attend.

Being welcomed by one's pets triggers happiness: the sight of a dog spinning with delight delights us in turn, but we humans yearn for a wider and wilder welcome as well, to be greeted across the species divide. Hence the joy people feel, even via a screen, at the sight of lions giving their human companions a rapturous hug. Yes, trained behaviour means animals may sometimes be seeking food rewards, but still the delight in us arises when animal mind seems to include humans. We seek an invitation into the world of fur and feathers and unfathomable eyes which see an infinity of worlds within this one, because the true, deep sense of life is formed of inter-mindedness.

If humans deliberately call for help from honeyguides, the birds will directly respond. Such communication between animals of different species is rare, but other examples include the relationship of the orphaned baby hippopotamus looked after by an Aldabra tortoise 130 years old, who vocalize together in neither typical-hippo nor typical-tortoise ways.

From childhood onwards, we have a poignant desire for animals to communicate with us: *The Jungle Books* are so appealing because Mowgli can talk with the animals and they with him. Koalas, like living teddy-bears, epitomize the sweet furry world that children and clear-hearted adults cherish: to have one wrapped in my arms, to stroke it, love it, protect it, hug it, press my nose to its nose – I don't know if anything in the world would make me happier. Child-mind leaps at such characters as Dr Dolittle and adores Aslan and is enchanted by the beguiling tales of orphaned humans brought up by animals: we hardly dare to believe they could be possible because we so want them to be true. It seems a visceral and innate wish, as if it confers an interspecies grace. Child-mind is devoted to animals, as children's authors and marketing departments know: children dream of animals, talk to them, are fascinated by them.

Communication between creatures and humans is a fixation of science and has led to curious discoveries: dolphins communicating with humans will modulate the pitch of their calls to stay within the realm of human hearing; orangutans will modify their gestural signals according to the comprehension of their human audience.

Such unfeigned communication, unbuyable and uncommandable, delights us as if they the unfallen were in that moment inviting us to step across, right through the curtains into the Dreaming. 'Everything has and tells a story. Everything communicates, through its own language and its own Law,' say Indigenous Australian Yolngu people from Bawaka in north-east Arnhem Land. Indigenous cultures have kept faith with the animals as part of what it means to belong, and the world is larger and more vivid when animals and birds and insects are imbued with spirit and significance, when there is Mind of unknowable diversity, elastic and ecstatic, until the very air is electric with Message and there are more stories than stars.

The communication between animals and humans is sometimes a terrible reproach. While elephants in captivity can speak human words, wild elephants have a word for 'human being' and, points out animal philosopher Eva Meijer, in *Animal Languages*, it indicates 'danger'. I have always wanted to hear a koala call. I have never wanted to hear one cry for help, its fur singed, its paws and nose burned, crying little bleats of bewilderment, and whimpering with pain in the arms of the Australian woman who rescued it from one of the bushfires caused by the climate crisis. Something in me died that day, and I'm not alone. We need their well-being, their voice, their happiness, their life.

When other creatures speak to us, a breach feels healed into wholeness, wellness. Worldwide, shamanic lore has included the art of shapeshifting; these animal-transformations are often treated as a fact without much analysis but the revelation to me is that healing, whether individual or social, is thought to come about through *animal* mind. Animals are the Healers, if we would but let them. This is physically true, as we know that, for example, heart surgery patients recover more quickly if they have a cat on their bed. Dogs can detect certain cancers through their heightened sense of smell and some dogs are now being trained to detect Covid-19. Emotionally, animals are the first-responders for the human heart, and eschewing the natural world is life-denying, refusing its most potent medicine: vitality.

Vitality is at the heart of healing traditions: acupuncture or yoga, the concepts of Chinese *Chi* or Indian *Prana*, the life force in flow. It is among the five 'character strengths' most correlated with happiness, according to the *Journal of Social and Clinical Psychology*, the others being curiosity, optimism, gratitude and the ability to love and be beloved. Vitality means living in vividness, alert, the senses picking up everything. It is the embodiment of life, keener and more alive. It is a core inner strength and not necessarily correlated with age: an eighty-year-old can be elastic with vitality. It is zest, enthusiasm, energy: sheer sap-rising, the very quick of life.

This place of flow is where the current runs fastest, the quickwater of a stream: and the flow-state is where animals dwell and, according to Rupert Sheldrake, this is why their company is healing: they draw us into their flow-state, which is essential for creativity and, in turn, happiness. Happiness includes being aware of the mindedness of others: it is energizing, and excites a curiosity towards difference and the unexpected. We become avid readers of multitudinous stories. Being intricately at home in the animal world increases the vividness of our experience, listening for prey, scanning for predator, smelling scents in the air.

Vitality is the aspect of human happiness that is most keenly associated with nature connection, as natural environments improve emotional functioning and attention. To notice, to attend the world, to be alive to its co-vitalizing, amounts to biophilia, the term used by biologist and naturalist Edward O. Wilson to describe that lovely innate quality of life loving life, and the particular kind of energy it offers is that shining momentness that, in the Homeric world, surrounds the gods: *energeia*. It is intense presence, wildness incarnate. In this sense, wild animals are the gods still walking – swimming, tumbling, climbing, pouncing – in the world.

On the second day of the trek in Sumatra, myself and a friend asked the guides if we could all walk quietly through the forest without shouting or disturbing the animals. That was when we saw them properly, undistressed, a mother and her tiny baby. The baby was peeking round the mother, curling its fingers to its original world, and dangling from one tiny arm. All is swing, on the ropes of liana. Tiny, fluffy and tawny, it could hang on its own arm as on a bungee rope, twirling 360 degrees one way and then untwirling the other way, ceaselessly elastic. The mother's hand was the length of the baby's whole arm, and she could scoop it up if she wanted to and rub its face, rolling her hand on its head. When the orangutans are happy, their happiness is infectious, with their huge, easy softness, dreaming, chewing, mulling. They looked at us, thoughtful, with those ancient eyes as they held hands in the slumbering afternoon.

Bhutan's Gross National Happiness Index contains nine domains, equally weighted, including ecological diversity and resilience. Happiness is, and must be, ecocratic. Our happiness depends on the happiness of all life: animals, birds, fish and insects, plants and trees. To each – the orangutan, termite, plankton, salamander or swan – its equal due, and an ecocratic sensibility is the absolute opposite of the celebrity culture or social media competitiveness that makes so many people feel so miserable, for ecocratic happiness does not run a *Scala Naturae*, does not care about status and 'Likes', and excludes no one and nothing.

Loneliness is, we know, a root cause of unhappiness, while a sense of community enlivens us and makes us happy. But there is more: interspecies community matters. Animals are vital for our happiness as an unlonely species. But this is the age of our solitude and many humans feel estranged from the world in this species-loneliness, outcast from the intensity of the fully thriving world. The non-human worlds of animals and insects are the only other life we know of in the universe: without them, how silent and foreboding is the loneliness of humanity.

Alone, together. Mass isolation. In the collective trauma that was the outbreak of Covid-19, many of us felt that appalling scythe of isolation cutting us to the quick. The loneliness of it shrieked through millions of lives. People sought out pets, bought dogs, found kittens. On one day of acute loneliness, I lay in a field with a foal a few days old. It nuzzled me, breathed into my nostrils, and I breathed into its, as I stroked its soft teddy-bear fur. Then it gently laid its front legs around my neck and shoulders and I lay there comforted, hugged by a foal.

In lockdown, the resurgence of wildlife was a consolation and a delight: dolphins on the Cooloola Coast in Australia, with fewer people than normal at a regular feeding station, showered humans with gifts of coral, bottles, anything to communicate and call to us. There were goats in a town centre in Wales. When asked their single favourite thing about lockdown, many people answered in one word: birdsong. It made us happy, it made our hearts sing in that exquisite but imprisoned spring. And yet what do we humans also do with our love for birdsong? We cage the songbirds. Song thrushes may be blinded to make them sing more. A red-hot needle in the eye, to make them sing for us.

Covid-19 is thought to have originated in the relationship between humans and animals, as John Vidal, in an important and early analysis of the outbreak, writes in *Scientific American*. It is an issue of planetary health, he shows, an issue of how the living world is treated. More than simply a medical issue, it is about the shared health of a shared home. It is a result of the savage plundering of the habitat and homes of the more-than-human world, the slaughter of wild animals, meaning that humans are in closer proximity, and diseases can jump across species. The so-called 'wet markets' may well be responsible, selling fresh meat where living animals are killed to order, sliced up and sold, including live wolf pups, pangolins, salamanders, crocodiles, scorpions, rats, squirrels, foxes, civets and turtles. But this is not the full story, Vidal shows. Zoonotic diseases, of which this is one, very seldom cross over to a different species, but mining, logging and all kinds of deforestation make it happen more easily. When landscapes are destroyed, the species that thrive in the changed locale often include bats and rats that pass on pathogens to humans more easily, and the trade in wild animals (one that includes the heartbreaking trade in caged birds) adds to the damage. Meanwhile, a tiger in the Bronx Zoo is infected with Covid-19. Caged, isolated, torn from its destroyed habitat, it sickens with a disease caught from an infected zookeeper.

As I write these sentences, the sense of lockdown-isolation hurts. Each of us separated from each other and never lonelier, feeling the loss of our relationships with other humans, while the wider relationship between humans and the entirety of the living world is what most needs healing, a toxic relationship that needs the rarest of medicines: a change of heart.

The philosophy of human exceptionalism, that arrogant and ultimately self-injurious idea that humans are a species both separate and superior, reaches its apogee in mass extinctions.

'Beware what you wish for' is the fairy-tale adage. Humans are exceptional in seeking to be excepted, and our wishes are coming true in a tsunami of extinctions. The human voice is becoming one monotone, as savage as it is sad; we are alienated from the healers, deaf to the singers, turning our face to the wall, a deadness at our hearts, lost to the vivid world, self-exiled and self-estranged in a sump of the spirit.

Life without the proliferation of creatures is like prose without metaphor, words without music, fact without dance, all the iridescence of a kingfisher reduced to dusty grey feathers in a dead-handed museum. All around us, exuberance is suffocated as life dims, unlustrous without animals, unlit by bird-flight. Tennyson's elms immemorial are gone and all but forgotten and his innumerable bees are facing their countdown. This is what the *ghast* means, an abyss of the unliving. These extinctions are the slaughter by an indifference too stupid even to be appalled, and we humans, instead of being the lucky clever ones, at home within the myriad beauties of the living world, are instead the pall creeping around a funeral of life.

In Mexico once, I heard a miraculous sound, the wings of thousands of butterflies applauding the sun. The sky streamed with them, the trees were orange with butterflies sheltering on the leaves till the sun warmed them to flight again. Monarch butterflies fly in their millions from North America to overwinter in the mountains of Mexico. Two thousand miles of butterfly flight, incomparably precious, which they have been travelling for millennia. The Monarch butterflies gather here in the world's centre, an *encuentro* in the mountains, rustling their wings like a million scripts for secret wishes in a whispering sky. How could you weigh such light or such lightness? Half a gram, hardly a breath to make a word or a note.

The ancient Greeks represented the goddess Psyche as a butterfly; it is a creature widely understood to symbolize the soul. 'The butterfly is the soul of nature. Where there are butterflies, there is goodness and strength in the natural world,' a local man told me. And this is the heavy-hearted part. Monarch butterflies feed on milkweed. Milkweed is being killed by Monsanto's herbicide Roundup. Genetically engineered corn produces pollen which kills Monarch butterfly larvae. Deforestation robs them of their shelter. As a result of all this, 970 million Monarch butterflies – 90 per cent of the total population – have disappeared across the USA. They begin their migration at the equinox, but if they are cheated at the equinox by weather which is unseasonably cold as a result of an already-changing climate, they will freeze to death.

While I was there, one alighted right by my boot, and such fragile beauty stops you in your tracks. No one in their right mind would stamp on one but, collectively, it seems we are choosing to kill them in such numbers that now the Monarch migration – one of the greatest migrations on Earth – is under threat and may itself be brought to an end.

A butterfly stamped under a boot. A damselfly torn like turquoise silk. A match struck and a bee wing set alight. While acts of wanton cruelty – for example the shooting of a giraffe by a Texan woman who gloated of being 'blessed' by this murder, or the shooting or stabbing of dolphins – may be held up to public shame, yet grief itself goes numb at the numbers as we collectively commit a bureaucratic slaughter that beggars any genocide.

It is the individual deaths that hurt the most, prompting anguish. Hares are rare where I live, and their flickering, electric life-force makes me gasp. One spring morning recently I found a still-warm corpse, a hare shot and left for the crows, the light only just gone from its eyes. In the last harsh winter, I saw a wren frozen to death, its two bright blackberry eyes now a blind wince of ice. It had fallen by the path at nightfall, fallen silent, a dying fall into oblivion and, through one bird, all the fallen rose to my mind, in an immensity of extinctions.

You know the figures. Wildlife populations have fallen 68 per cent since 1970. We have killed them directly or indirectly. Since 1980, populations of common birds in Europe have been slashed by 421 million. In the UK, the number of nightingales fell 91 per cent between 1967 and 2007. Puffins are suffering a grave and mysterious population decline. Almost 38 per cent of all known species are on the edge of extinction, because humans are killing species at between 1,000 and 10,000 times the natural extinction rate. Puerto Rico has lost 98 per cent of its ground insects. Forty per cent of the world's insect species could be extinct within a few decades. There is likely to be a consequent explosion of extinctions because each creature is so sweetly tendrilled into so many others and no species can live alone.

I see him cloaked in cold mist, the Fisher King, a desolate figure in a wasteland of his own making. Wounded, he hunches silently over his fishing rod in a sea of silence, catching nothing except the reflection of his own shroud in the dead water. He is us.

The Fisher King, keeper of the Holy Grail (the cup of the Last Supper), is an enigmatic figure in literature: a rich king wounded by his own spear. The earliest sources show him as suffering a moral wounding – a result not of accident but of his own ethical failings. The wound does not heal. Worse, its effects creep out, killing everything around him so the abundance and richness of life is reduced to barren waste.

There is a moment in the story when the wound – of both the king and his environment – could be healed, not by medicine or technology but by intelligence. The young knight Perceval arrives and has the opportunity (which he misses) to pose the right question to the king.

The healing question, the timely question, is this: Whom does the grail serve?

The Fisher King is emblematic of this Age of Extinction. For millennia, the oceans have thrived with stupendous life, from fish in kinetic rainbows racing currents to coral cascading colour. Yet human activity over the last thirty years has poisoned the oceans and exhausted the seas, turning this blue world into a dead sump of denied life.

Humanity has manufactured a marine wasteland: part of the sixth mass extinction. The first to be knowingly caused. Two thirds of the species we have fished since the 1950s have collapsed. Some species are down 99 per cent. The oceans, says Professor Callum Roberts in *Ocean of Life*, have changed more in the last thirty years than in all of human history before. The sea is suffering a sea change into something bleak and strange.

Most people will never see this underwater world and know it only when its meaning has been processed into money, when its beauty – as surprising as an octopus's garden, as mysterious as whale song, as appealing as a clownfish – is merely a discarded by-catch.

Like a collective Fisher King, modernity is wounded by its own aesthetic failure to see beauty if it can grasp profit, its ethical failure to register the value of life itself if it can register economic advantage. Technology has turned us into kings, but intelligent ethics has not kept pace, and humanity itself will suffer from this self-wounding.

Although such a devastating stupidity has been unleashed on them, the oceans themselves are associated with deep wisdom. Whales and dolphins are known to possess enormous intelligence, and the oceans have long symbolized depths of thought and immersive insight. There is something unhallowed in the extinctions we cause knowingly, a devastation of the soul when the holy grail of profit is pursued at the cost of life itself.

The deadened oceans are an analogy for a wasteland of human imagination, while both mind and ocean should be generous in generation, with teeming, shoaling, spawning, frothing, helical life spiralling up from the depths, bubbling at the top and spinning back down again, that liveliness of life, the vitality of ocean and intellect on which humanity ultimately depends. This life has been shrouded, polluted and injured, and the public mind barely notices. By contrast, the rise of profits, the 'health' of the economy, is treated as if money were somehow more alive than life. The artificial world more prized than the real one.

Industrial fishing has meant that, since 1970, the numbers of bluefin tuna have declined by two thirds. At a world conference of CITES (Convention on International Trade in Endangered Species) in 2010, there was a call to list the bluefin. Japan lobbied against it and won. And then the Japanese delegation served a grail supper, a last supper, a banquet – of bluefin tuna. There is a terrible, toxic poetry to this: when profit is the holy grail, ours has become the age of the Fisher King, waiting, self-wounded and devastating the waters, waiting, waiting for the question that modernity has so far failed to ask itself: Whom does this grail serve?

When the meta-study that reported on the collapse of insect life, and therefore our lethal bequest to our own children, was published, it was poked about a bit for a few hours in the media mind. When, a few days later, Karl Lagerfeld bequeathed £150 million to his celebrity cat, the media went wild for the story. One pampered pet, now the richest cat on the planet, was more important to the media than the most grievous threat to life. Human voices are drowning all other voices. We pass through the world, we the misguided ones, leaving a swathe of silence because of a deadly lack of imagination. Animals, pure embodiments of vitality, are being erased from the Earth as fish are from the ocean.

The language we use for this is itself deadly. The mass of ocean writing is a heap of broken plastic words: stock, fisheries, industry, off-shore, tonnage, commercial fleets, sea cages, fish farms, subsidies. Through that language it is hard to see the ocean's true nature, whose vitality needs to be rendered as beautiful as iridescence itself. We speak of an 'extinction event' or 'species decline' because of 'intensive agriculture'. These are lifeless phrases. How easily the eye bypasses them. They are words of tarmac and traffic, not the lovely writhy ivy words of the woods.

I cannot touch or taste terms like 'habitat loss' or 'pollution' because they are unbeloved words which carry within themselves the toxicity of lifelessness. Humans, we are told, need insects for 'the function and services they provide'. Cold language, cold as coins on corpse eyes, cold as the philosophy that put us here. Words of heart are needed. Truth and rage, as the movement Extinction Rebellion articulates. Because this is a crisis of deadly stupidity like no other. No war matches it. No genocide. And there is nothing greater than to be in service to this, the life on which any chance of a happy future depends. Being alive in this age puts on us all a mantle of responsibility of a wholly different order.

There is a new word in the air: *defaunation*: the loss of absolute animalness. Defaunation includes the loss of individuals and the loss of abundance. Defaunation, argue researchers in *Science* magazine, should be as familiar and influential as the word 'deforestation'. Another term for the loss of the world's wild fauna is 'biological annihilation'.

Please tell me you understand the immensity of this. And if you don't, please think, alone and quietly perhaps, of the unfolding ending. Let me speak simply into the simplicity of your heart, then, and let me just ask you what you love, what makes you happy.

Is it a child? Is it your partner? Do you love your friend or, Little Prince, do you love your rose? Do you love your dog, your cats, your church, your home, your garden?

Your books, perhaps, or the poetry you make, or the music? The meaning you have made of your life, maybe, your health, status, honour or all of these? And this love, then, this happiness that you hold so dear, tell me how it will even exist without the tiniest of beings, the insects, against which we have been so utterly pitiless? Without the insects for the food and the flowers and the soil?

Insecticides should be made illegal overnight. Every scrap of land turned to organic agriculture. Every shred of mental energy requisitioned for love, essentially, the love of life.

This is the end, my friend.

And this is the age of the endlings, the very last individuals of their species. That word, *endling*, resounds its requiem like every minor chord that has ever been played. In 2012, Lonesome George, the last Pinta Island tortoise, died in the Galápagos Islands and the species became extinct. The chestnut clearwing moth is an example of an endling from the insect world. Words with that suffix *-ling* are diminutive, tender, poignant: foundling, and starveling. They are often bird-related: nestling, and fledgling. The endling of a subspecies of rufous-fronted laughing thrush, a little female bird called Esa, 'the lonely one', is looked after in Java, but at her death a world ends. I will write for her.

To hold in your mind an endling, to stroke its feathers, feel the pulse of its heart, listen to its last and dying song and then to watch the light grow dim and fade and see it fall, then, the last ever, its light put out and knowing there will never be another.

The neverness. The chill at the bone and no grief like it.

And yet, and yet, and yet, it is also true that there is a honeyguide in the human heart that knows its true north. Children have it, they who still dream of animals, who talk to them and for them, which is why they are speaking as the consciences of the world: twenty-one children suing the US government over climate change; youth-led climate strikes spreading worldwide; teenage activist Greta Thunberg, one of millions. The knowledge of what we have done to the natural world is – at last – causing us the grief and shock that can be transformative and lead to change. Within each of us must be found *energeia*, and the quality of life loving life sufficient to rebel against its extinguishing, to say 'never' to the toxic tragedy that brought us here and, instead, seek happiness only when it is ecocratic, made of chlorophyll and birdsong, the laughing generosity of macaques leaping through the trees.

I went swimming in a lake near me yesterday. The sadness of extinctions was sweeping through me. And then I saw a water boatman who made me smile and I am glad that I can still be comforted by the most ordinary of insects rowing its extraordinary message of metaphor: keep on rowing, my friend, keep rowing for your own far shores.

The Narcissist and the Firefly

Imagine if our food were brought to us by dedicated and almost invisible angels. Imagine them flying, effortless and iridescent, with a beauty more extraordinary than any art of ours can ever replicate. Imagine if those mysterious beings worked freely to keep alive almost the entire living world, including birds, animals and ourselves, offering us a myriad of flowers and the feast on feast of exuberant life by which each of us is heiress to the bequest of millennia. Imagine if these angels also gently and tactfully disposed of the dead, unobtrusively cleaning corpses, tucking the dead back into a deep bed of earth so they can rebecome life in another form. Quiet and kind, these angels remove death by increments and, without them, dead creatures would exist for ever and we would wade through corpses with every step we took.

They do not take the title of angels, being by nature bashful and unassuming: they go by other names: firefly, bee, ant, caddisfly. We humans, it seems, value irreal angels more than the priceless reality of insects.

A secret commonwealth, the insect realm

encompasses more species than we have identified, though the known ones include a plethora of life: the katydid and cicada, the grasshopper and froghopper, the mayfly, cranefly, the green lacewing and antlion, the violin beetle, giraffe weevil and the sunset moth of Madagascar, a day-flying moth whose wings are rainbows. The insects: hallowed be thy names, some of which are pure poetry: the orchid bee, coloured in bronze and ultramarine, purple and gold; the ladybird; the glasswing butterfly and the emerald swallowtail.

Insects may be invisible in the dark, like fireflies with their almost magical glints of bioluminescence glimpsed amid the blackness, glowing a moment, then gone. *Another! Just there! Look!* Insects are often secluded in foliage, or, disguised by their very numbers, they may appear like a haze, in a gauze of translucent wings. Hidden in their very tininess, they are, together, a gigantic collective of goodness, dancing in constant attendance to living things. The insects pollinate three quarters of our food crops and 80 per cent of wild-flowering plants and keep the soil healthy, recycling nutrients. From their actions flow the countless forms of life, from the apple blossom to bread and roses and the silver salmon, indeed everything that has ever flowered and ever will. And from physical life flows everything to be treasured in human life, from existence itself to the highest of the arts. It is ultimately thanks to the insects that we owe the flowering of plants and the flowering of culture alike. Michelangelo salutes them. Notre-Dame bows in thanks. No insects, no Mozart. Without the beetles we would have no Beatles.

The insects are music of themselves, in an orchestra of crickets or cicadas. They give us choirs of insects, flights of insects sing thee to thy rest. Without them, there would be no chorus of frogs or birdsong: for how would a bird sing without insects to give it life and music? To them we owe everything: my life and yours and yours and yours. According to an Indian proverb, 'As day breaks, the glow-worms say, *We lit up the world!*' And they do, all the insects, lighters, delighters, makers and sustainers of worlds. The rusty brown and white Atlas moth, with a wingspan of twelve inches, seems named as an emblem for the whole insect realm: they truly carry the world on their shoulders.

In summer, swimming in a lake near me, azure damselflies are a thread of sky that stitches me into the tapestry of heaven. I mean it. I am lost in turquoise adoration: the flight of a damselfly is the ever-uncatchable quintessence of dream. As nymphs they may spend five years underwater before climbing up a plant stalk and into the air one night, and metamorphosing into an adult.

In the air soaring above my head as I swim, or swooping down to my eye level, swifts are at their arabesques, catching tiny insects. The very sky is alive because the flights of insects in turn enable the flights of birds. But the air, once abundant with insects, becomes enemy territory for many birds now, a migration route an ordeal across deserts, flying over vast fields of industrial-agriculture, for without insects to feed on, the birds risk death at every wingbeat through the starving skies.

Imagining a world without wings fills me with inconsolable sorrow. A wren, hungry and songless; a swift dropping to its death; the air emptied of life. Without insects and birds, we rob ourselves of all that flight represents: the wings of mind, the flight of imagination, that mother of empathy.

Meanwhile, hunched on the banks, Narcissus-modernity leans over the pool to gaze at his own reflection. He misses the nymph, the dragonfly and the swift. He cannot see the flights of insects in the air. His mind is grounded. Narcissus knows no wings, feels no empathy. His narcissism makes him cruel because he sees only himself. Disdainful to those who love him, he finds no mate, creates no future generations, and fades away, his body disappearing, until all that is left is a little flower. He dies, eventually, because he is able to love only himself and no one is alive in his heart. None so alone. In myth, what strikes me most about Narcissus is this isolation. Gazing in deadly fascination at himself, he is, in the end, his own worst enemy. Insects have antennae and compound eyes for sensitivity and vision: Narcissus has neither.

Some societies have long refused the idea of human isolation and have cherished the vital relationship to every part of life, where even the tiniest being is necessary and sacred. In Costa Rica, a suburb of San José has given citizenship to every bee, bat, hummingbird and butterfly. Devotees of Jainism practise *ahimsa* – non-violence or harmlessness – to the degree that they sweep the ground before them as they walk to avoid treading on insects, which is why the insect mythologist Ron Cherry calls Jainism the most 'insect-friendly' religion. Japan used to be called Akitsushima, Land of Dragonflies, which are taken as symbols of courage, strength and happiness, appearing throughout Japanese art and literature. The Albanian language has two verbs for dying. One, *ngordh*, is used when any animal dies. Another verb, *vdes*, is used only for the death of a human being or a bee, so close is the association between the sweet bee and human life. Bees were honoured by being depicted on ancient Greek coins from Ephesus long before it was known that a hive of bees has to fly the equivalent of twice around the world to create a pound of honey.

The oldest continuously surviving human cultures – Indigenous Australian civilizations – respect and honour the tiniest creatures in a kind of ecodeistic reverence. Honeybees play a significant role in funeral ceremonies for the Marrangu people. In central Australia, honeybees feature in totems, songs, rock art and creation stories: in the beginning was the bee. Across Australia, insects are honoured in songs and feature in mythology, place names and personal names. Insects are part of rituals, including the 'increase ceremonies' once widely held to promote the abundance of particular species.

Insects are thought to have been alive on Earth 480 million years ago. That's 1,524 times longer than us, as *Homo sapiens* has been here only 315,000 years. Insects who were there long before our beginning, and have been cherished by humans for thousands of years, could vanish within a century, a whole world quietly disappearing.

In 2020, a meta-analysis published in the journal *Science* showed that land insects were declining at the rate of about 9 per cent per decade. A 2019 study published in *Biological Conservation* showed a 2.5 per cent rate of annual loss over the last 25–30 years. In ten years, at that rate, there could be a quarter less insects; in fifty years only half would be left; and in a hundred years, none. Some insects are on the increase, including locusts, cockroaches and horseflies, but the insects most badly affected include bees, butterflies, moths, beetles, dragonflies and damselflies. Light pollution (as well as habitat lost and pesticides) endangers fireflies' existence, because their courtship happens in the dark: they can find their mates only by their guiding lights. Meanwhile, when other insects, such as moths, flock to artificial light, a third of them will be dead by morning, either being predated upon, or simply dying of exhaustion. The number of widespread butterfly species fell by 58 per cent on farmland in England between 2000 and 2009. The reasons include habitat destruction, the climate crisis and the heavy use of insecticides.

Organically farmed lands do not show the same devastating decline of insects. Moreover, not every part of the world shows such declines, for the simple reason that similar studies have not been done across the world. The studies that have been done have been occasionally criticized for not showing a pre-pesticide baseline for the numbers of insects. Why did they not show that baseline? Because such studies simply do not exist. The insect realm had once seemed as infinite as the numberless stars, a myriad of proliferating life. So abundant were they that no one thought to count such countlessness or measure such multitudes.

Recent studies are clear: gone is the windshield phenomenon whereby, driving in summer, it would be necessary to stop frequently to clean the insects from the windscreen, while the 'moth snowstorm', a blizzard of moths in the headlights, was the experience of every night drive. Surveys of insects hitting car windscreens show a 50 per cent decline in the UK between 2004 and 2019, and an 80 per cent decline between 1997 and 2017 in Denmark, with a consequent decline in the number of swallows and martins that survived on the insects. Eight in ten partridges have vanished from French farmlands. Nightingale numbers have dropped 50 per cent and turtledoves by 80 per cent. Half of all farmland birds in Europe have gone in thirty years. The birds, winged messengers, can read the news and know it for grievous truth, and they are flying starving into the Great Silence. And we call ourselves sapient.

It was the studies of insect collapse reported late in

2018 that first made me cry for insects. The horror of it swept over me: I cried for three days. I hate all kinds of bullying, and the fact that the insects are the tiniest creatures, bullied by humans acting as monsters, gave the facts an edge of very personal pain. It was of course infinitely more than this: I saw in one awful moment a vision of the desolated world, a devastated wasteland.

Writers sometimes tell their readers when they struggle for words, when they experience writer's block or when their psyches demand a fallow period. That admission is a touching one, a truth so precious that I do not use it lightly. I use it now. The magnitude of this situation silences me. The words I lean towards are not enough. Tears, maybe. The raw scream of rage and pity, perhaps. But what words do you suggest I use here? Annihilation? The end of worlds? The last generation? Absolute apocalypse? If you were looking this full in the face, what expresses it sufficiently? And a savage anger overcomes me. This is not a game. *Nature is not a hobby.* Protecting insects, children, and life itself is a necessary duty, incumbent on us all.

With the insects gone, we would lose the birds, reptiles and snakes: the gold day dust gecko, and the thorny dragon, the sea turtle and the rainbow boa, iridescent and nocturnal. What is the point of listing what would be lost? We would lose, to put it bluntly, almost everything. Starvation would stalk the land for almost every kind of creature, including ourselves.

I don't want to be lyrical now. I just want to swear.

The collective stupidity renders all my craft useless. What writer's art can ever convey the vast, deadly and deliberate slaughter, with all its consequences that are, in sum, the sum of it all. The Everything. Where to go with this gigantic stupidity? What the fuck did we think we were doing? Why the fuck are we still doing it? Industrial-scale intensive agriculture is killing ecosystems by killing the insects. If I sent a tweet, I would write only this: *Mass use of insecticides leads to mass death of insects. And I'm, like, DUH? Who knew? FFS.*

What can you and I do, right now? Heed the climate crisis. Do things that seem small but are not: turn off outdoor lights at night. Only ever, ever eat organic food. I beg you. I implore myself. And if – when – you say you cannot afford it, I understand. Then speak about it. Do not be silent. Tell everyone around you. Tell everyone in power. Take it to your council. Vote without compromise for green politicians. Sit outside the office of your political representative and don't leave. Anyone who really allows themselves to take in the magnitude of this ending, the threat to the very existence of us, feels the weight of world-sorrow. Others felt it, too, and a friend of mine passed me a tatty leaflet. 'Please give this to Jay,' someone had said to her at a festival: 'she will like this.' Never has one little leaflet made such an impression on me. Two words: Extinction and Rebellion.

We are living in mythic times, and not in a good way. This is the age of narcissism. We are being governed by narcissists. Key characteristics of narcissism include entitlement; an exaggerated and unearned sense of self-importance and superiority; grandiosity; and an excessive need for admiration. Narcissists lack empathy; disregard others' feelings; expect compliance from others; seek those who will kowtow and serve; cannot tolerate criticism and suffer envy. Populist candidates score high on narcissism traits. I see Trump offering to buy Greenland. Johnson proroguing Parliament lest it foil him. In Turkey, Erdoğan destroys a forest reserve to build a $600 million palace for himself. Putin rides half naked on horseback as he undermines democracy in the USA and UK as well as in Russia. In Brazil, Bolsonaro; in Italy, Matteo Salvini; in the Philippines, Rodrigo Duterte; in Hungary, Viktor Orbán.

Collective narcissism includes an unrealistic but rigidly held belief about the in-group's greatness, believing that the in-group is special and uniquely endowed. There is a tendency to aggress against out-groups and a swiftness to perceive threat and insult from them. Politically, collective narcissism sits well with right-wing authoritarianism, a preference for military aggression and blind patriotism. It is related to ethnocentrism and a desire for national supremacy – White supremacy, in Trump's case.

Perhaps the most significant historical example of collective narcissism was Nazi Germany, where the narcissistic (and indeed psychopathic) Hitler, with endless appeals to nationalism, German exceptionalism, right-wing authoritarianism and militarism, chose the Golden

Aryan Child while also creating the Jewish Scapegoat out-group, describing Jews in the pernicious language of vermin. Pests. The tiniest of creatures, the insects.

We are not only governed by narcissists but are living in an age when narcissism is on the rise. In *The Narcissism Epidemic: Living in the Age of Entitlement*, the authors Jean M. Twenge and W. Keith Campbell consider that the USA is suffering an epidemic of narcissism, as narcissistic personality traits rose as fast as obesity from the 1980s to 2010. Consumerism and easy credit encourage the endless sense of entitlement that is so characteristic of narcissism. Advertising offers the glib but terrible narcissistic justification: *Because you're worth it.* Celebrity culture lures people to a peculiarly narcissistic triumph of vanity and self-admiration, attention-seeking without concomitant accomplishment. The Internet operates like a vast pool in which anyone can play Narcissus, gazing into their Facebook profiles and Instagram accounts, contemplating their own image in grandiose fantasy, counting their superficial 'Friends' and shallow 'Likes'. Meanwhile, as Twenge and Campbell report, children are encouraged to feel that they are special and superior: one preschool in Manhattan designated September as 'All About Me Month'. One lesson-plan, for two-year-olds, spelt it out: 'Today we will study ourselves *in a mirror.*'

The post-war years have also seen the rise of what I would call Species Narcissism. Of course, the idea of human exceptionalism and superiority to the entire non-human world has a long history, but the tools to enact that worldview, the machinery and chemical warfare, have multiplied in lethalness in the post-war years. The in-group is humans. The out-group is the rest of nature. Yes, there are a few granted favour: charismatic mega-fauna and domestic pets. But for multitudes of creatures, their experience of life, whether it is in factory farming, animal cruelty or the loss of habitat and home, is unremitting suffering. 'In their behavior toward creatures, all men are Nazis,' wrote Jewish author Isaac Bashevis Singer, who fled his native Poland fearing the rise of the Nazis. 'For the animals, it is an eternal Treblinka.'

The insects are the most despised of all creatures, the ones killed in the largest numbers, even though, with every litre of insecticide, we are threatening our own survival without noticing. Sometimes the worth of the insects is measured in our money. 'Eco-system services', we say, as if the living world exists solely to serve narcissistic humanity. The insects that pollinate some three quarters of human food crops are said to perform a 'service' worth perhaps $500 billion every year. Insects deliver to *us* these benefits, costed into *our* dollars. Insects are beautiful to *our* aesthetic. This is the particular tone in the dialogue that surrounds narcissists. They – think Donald Trump – need everything to be related to their own self-interest. They require others to pander to them, servile and grovelling: if others do not please and placate the narcissist, they will be rejected and punished. So the more-than-human world must appeal to our narcissism, taking that obsequious position: they are there to serve us. The narcissist exploits others without care or respect. Just so, the relationship between modernity and the rest of nature is overwhelmingly transactional.

Without the insects we, like Narcissus, will die of hunger. Why? Because modernity believes that money is more important than the living world, that the profit-motive comes before everything. Even the children.

Very few parents disinherit their children. One group, however, frequently does so: narcissists.

Narcissist parents often choose a Golden Child, who are themselves often walking the narcissistic path, and they give that child everything. Others are made into Scapegoat figures, rejected and disinherited, of no value in their parents' eyes. We see it in the Trump family, where Donald was the Golden Child and he coerced his father into removing his elder brother's children from the will. (They were disinherited: they sued: they won.) This is what I see happening collectively now. Modernity, suffering malignant narcissism, is disinheriting its children en masse, except for a few Golden Children who have a vast inheritance, as wealth disparities favour an ever-smaller minority: the world's twenty-two richest men are wealthier than all the women of Africa put together. (The young, seeing themselves disinherited, are trying to sue for their right to live. May they win.)

The sweet cascade of generosity, by which generations

downstream are bequeathed life and happiness, is being brought to an end. Modernity inherited the largesse of the Earth, much of it the gifts of those insects – food, silk, honey, flowers – but in just a few decades it has swarmed across the Earth consuming everything, a locust generation that leaves nothing but husks and carcasses for the children. Receiving the inheritance of angels, this generation disinherits its children, bequeathing an empty plate and a flowerless jar, and an abattoir of the ever-undead. 'Here you are, my darling, I wish you an abject birthday and a necklace of barbed wire to remember me by.'

The profit-obsessed mindset that has caused such devastation associates what is 'precious' with what is 'rare'. In the case of insects, the truth is quite the opposite. What is ordinary is precious. What is common is priceless. The secret commonwealth of insects is our true wealth. They are the gift-bearers, the common-or-garden cherishers of generations past, passing and to come.

In place of the ceremonies of increase, chemical companies dole out deadly rituals of killing. In place of creation stories that put the insects at the heart of life, every advert for insecticides is an annihilation story. A pesticide-sprayed field poses as food: it is actually starvation. It pretends to be profit: it is actually loss. It pretends to have killed only the insects: it actually claims the lives of the grandchildren. When bees are subjected to neonicotinoids (neurotoxins), they die by disappearing, so hive after hive is silent, still and mysteriously empty. Why? Because the bees cannot find their way home. I hardly know a sadder image. And when the bees die so do we, as our lives are intertwined. We will die together, in Albanian.

E. O. Wilson wrote: 'If we were to wipe out insects alone on this planet, the rest of life and humanity with it would mostly disappear from the land. Within a few months.' To be fully human is to have a moral imagination to see the consequences of what we do. Yet the extinguishing of so much of life is almost a silent subject. We discuss so many things that do not matter, but not something which matters on this appalling scale.

Charles Darwin, as a teenager, once wrote to his cousin, saying: 'I am dying by inches, from not having anybody to talk to about insects.' Now all of us are dying by inches from not talking about the insects. The bee-loud glade is hushed, and only in that final silence will we hear the echo of ourselves and know it by its hollowness. Fireflies are winking in the darkness, fireflies who once proliferated like stars in sheer, intoxicating plenty. One by one, their lights go out. Modernity's life on Earth, brief as a firefly, leaves a toxic dark for ever. In the last of nights, the last star put out on Earth, in the horror of annihilation's finality, a last firefly, lonely to its core in a vast wasteland, glows one last time and is gone for good. Perhaps it will be only in that kind of darkness that we could have seen the magnitude of what we have done and all that we have lost: a world, each other, and ourselves.

Coral's Swan Song

Have I ever misread something so terribly?

I thought it was beauty.

A while ago, snorkelling in shallow waters in Indonesia, I saw some pieces of coral glowing electric blue, a colour so strong it filled the waters with a fluorescent melody, a ringing blue that sang itself out – out – to the realm of ultramarine. I had swum across swathes of dead coral that day, and was shocked and saddened. This blue coral at least seemed vibrantly alive and possessed of an utterly ethereal beauty.

Over a decade ago, I trained as a diver so that I could write about coral reefs in my book *Wild: An Elemental Journey*, but in no dive back then had I seen coral of this pulsing glow, lit from within, luminous as a bluebell wood at twilight.

What stays in my mind from a decade ago is the colours of a coral reef. Here small fish, anthias, play a yellow scherzo, there the orange of an anemone fish shines out. The blue and yellow of the surgeonfish is like laughter across the reef and the parrotfish shines like a paradise of gold and turquoise. The fire dartfish zooms into view, its body all the colours of flame from its pale yellow head to tawny embers at its tail.

Coral looks like ferns and reindeer horns, like frosted trees and feathery fireworks, like fans of gold and white lattices; while whip coral looks like an ancient anchor, ropes of coral look like necklaces made of moss crushed with diamonds, sapphire and shells. The coral reefs I saw ten years ago made me think of culture as well as nature: renditions of civilizations: the patterning of Islamic art, a hint of Borobudur, a quote from the Pyramids, or pointillism or modernist pottery, they create their ornate architecture in trellised balconies and stupas.

Nothing expresses vitality like a coral reef. Angelfish look like they invented iridescence, their fins trailing the glory of it. In vivid proliferation, in this world of reckless beauty, life is lived only in rainbows. For thousands of years, the coral world has existed mostly unseen by humans, in a phantasmagoria of psychedelic dreaming. It is now spiny, now prickly: coral may be fiddle-headed or pronged, spirally, whorled, tubular, gauzy or gossamery. It appears inexhaustible in its profusion, a kaleidoscope, a wonder, laughing with sunlight, swaying and thriving in the sheer *ivresse* of life. A coral reef, I felt, was one of the places where the quick of things could be felt in every cell of my body, every neurone of my brain.

Coral reefs are as necessary as they are beautiful. Fish come here to spawn, as the reef offers protection for eggs. Sea mammals, including dugongs, raise their young on the reefs. Various medicines, including prostaglandin, come from coral reefs. Reefs are home to a third of the species of the sea, and are entire communities of inter-thriving life. Coral acts symbiotically with algae that lives in coral tissue, and the algae photosynthesizes, turning sunlight into food for the coral.

In the summer of 2015, more than two billion corals lived in the Great Barrier Reef. Half have now been killed, largely due to the crisis of climate change and the overheating of the oceans. Ninety-nine per cent of the Great Barrier Reef suffers some level of bleaching. Twenty-nine per cent of it died in 2016. A 2017 UNESCO report found that bleaching had impacted seventy-two per cent of World Heritage listed reefs. Based on current trends, bleaching will kill most of the world's coral within thirty years and, because it is a fundamental part of the ecosystem, the death of coral will cause a terrible collapse in the wider life of the oceans.

When the oceans become too hot, the coral expels its algae, and the result is first bleaching, then death. Without algae, the coral starves. Further, nothing thrives around a famished reef. In the endless expanse of deadness, on that ghost-reef recently, the only fish I could see were grey, eerily translucent, camouflaged in this colourless coral, hanging listless in the deadening waters. All the colourful reef fish I saw a decade ago – the fish like wishes in their brilliant and shimmery shoals, their gold and purple swift as psychedelic electricity – can no longer find camouflage in the ashen coral, and have fled.

When coral bleaches, it can go white in a couple of weeks, as if enacting a poignant image of grief and shock. Of this coral are bones made. Broken. Lifeless. Skeletons. The endless grey of pallid, lifeless ashes, sunk into dust. This is what I was looking at a month ago, when suddenly, in the midst of the devastation, I saw that unforgettable luminous blue, fluorescing and electric. I thought I was looking at beauty. I wasn't. When I got home and researched it, I found that I had been looking, unknowingly, at death foretold. Not looking at beauty but at tragedy played in blue.

When coral gets too hot, it produces a chemical to try to protect itself from the heat, and this is what makes it luminesce in this unearthly way for a mysterious moment. It still lives but will die, and in this moment between, it shines. It is the coral's swan song. Richard Vevers, creator of the film *Chasing Coral*, describes it as 'the most beautiful transformation in nature. The incredibly beautiful phase of – death. It feels as if the corals are saying *Look at me. Please notice*. This is one of the rarest events of nature happening and everyone's just oblivious to it.'

When you dive deeply, colours disappear. As you

first dip underwater, skeins of fluent light surround you, but diving deeper the light fades and colours disappear one by one. The rays of reds are gone by 200 or 300 feet. The rainbow is inexorably reduced. After the yellows and greens give out, only blues and violets remain. Then these in turn give way to ultraviolet, the colour we cannot see. The ultraviolet reaches deepest of all, beyond our sight, as the prefix 'ultra' means beyond.

'Ultramarine' means beyond the sea, and watching this blue coral glow in its moment between life and death, the ultramarine has a timbre all its own in the deepening. Blue is the colour of both grief and love. Blue fathoms the past, and is the colour of eternity. Gazing at this coral, in the indigo, lustrous and low, what I was seeing was the time between its life and its death, the blue hour, the last colour you see before the abyssal depth, an ultramarine sung in the key of dusk, radiant into a world that ignores it, and sinking, ultimately, into invisible and anguishing ultraviolet, an ultra-dying at an ultra-twilight.

This England

In the deserts, a woman is staring at the parched landscape of arid plains. Her heart is thirsty, she is downcast and homesick, yearning for the fertile lands where she was born, for its moist and tumbling leaves, for the cool mountains and rivers and the meadows of sheer and vivid green.

To allay her nostalgia, her husband builds colonnades and arches, fills a garden with plants and irrigates it all with a filigree of waterlines until it flourishes with myrtles, until almonds bask in the sun, until date palms sweeten and pomegranates swell, until grapevines curl their tendrils to the touch of his beloved queen.

So Nebuchadnezzar created the Hanging Gardens of Babylon for Queen Amytis and so, too, an ancient portrait of homesickness was carved, that love of one's land so specific that its loss can make you sicken for the place of your belonging, its exact contour and climate, where the heart finds its hearth. The raw imperative of a first love. Earth is home.

It is as if humans are born with the capacity to love the land on which we first set eyes. It is as if we have an inner template for this home-land-love, which is then adapted to the precise landscape of our childhood. This is a parallel for Chomsky's theory by which a child is born with the ability to learn Language, an innate, pre-set template which is then tuned to the specific language which surrounds them. So perhaps we are born with an innate capacity to love our land, though that may be riverlands of wild garlic and bluebells, or desertlands of catfish and hot springs; lands which talk with the brogue of heather or the vernacular of oak; lands which speak the argot of snow or the dialect of the savannah. For, despite the diversity of landscapes where humans have dwelled, the one constant is the ready love in the human heart.

In this, more than anything, is a demonstration of the indigenous human being, where the foot is snug to the land it walks, the language entwined with locale, a dwelling sung with desire.

When I wrote *Wild: An Elemental Journey* I was intrigued by many conversations with Indigenous people, from shamans of the Amazon to Inuit people of the Arctic, about their love of land, its wildness and its sense of home. Indigenous people said repeatedly, 'We are the land.' The land quickens those who dwell on it, and they quicken the land in turn, bringing it into deeper life. Exile from their lands – through its destruction or through land theft – ruptures their identity so painfully that it causes a sickness of the heart, mind and body.

This is a birthright, this love of land. This Indigenous

nativity is a profound aspect of our human identity. But for the English, it is a contorted feeling, knotted with nastiness and silence, complicated by racism, guilt and empire. For many people, it is also the source of an almost unfathomable nostalgia, an ache of the heart which all the hanging gardens of the world cannot console. The Brexit vote had complicated causes, including a measure of anger against the effects of austerity, a certain amount of xenophobia and a large measure of Russian interference. But one slenderly proffered cause has beauty: a yearning for something inchoate, almost inexpressible but emotionally powerful: a love of land.

It didn't come out like that. Instead of a discourse of belonging, the racist wing of Leave campaigners created the discourse of 'You Don't Belong' and levelled it at anyone it regarded as 'Other'. So a St George's flag was emblazoned with the words 'Refugees Not Welcome'. In Huntingdon, notices saying 'No more Polish vermin' were posted through letterboxes. Simon Woolley, founder and director of Operation Black Vote, comments: 'The Brexiters, with their jingoistic rhetoric, have put the country on a war footing. By framing the debate as "we want our country back", they have made immigrants the enemy and occupiers who need to be expelled.'

Belonging, from Proto Indo-European *dlong*, meaning 'long', gives us 'belong' and 'linger' and the yearning 'to long for'. It is a horizontal word, it lies down and snuggles, and has a sort of fidelity to it, a familiar, homely word, but one which has, to my ear, been sharply woken, dressed recently in nasty clothing, with a brief to expel and exclude. To kill. When Thomas Mair wished to symbolize his love of land and enact his sense of belonging, he chose to assassinate Jo Cox, MP, because she was a Remain supporter and he was a Nazi sympathizer.

The English can be poisonously racist towards Indigenous communities which know their cultural identity: they can also be envious of them. The envious response invites questions. Where are our old gods and songlines? The *genius loci*? The spirit of place? Why don't we, or can't we, sing our own folk songs? Why are the English so obsessed with selected pockets of history and simultaneously ignorant of most of it? What are the causes of our exile?

If you are a Palestinian, the cause of your exile is only too knowable, as poet Mahmoud Darwish discovered when he was six and the Israeli army destroyed his village, leaving ruins. He would become an 'internal refugee', a 'present-absent alien'. In 2001, Israeli bulldozers ripped open his village's cemetery for a road, churning up human remains, bulldozing the past. There are documented cases of settlers deliberately spraying raw human sewage over Palestinian homes in Abu Dis, a suburb of Jerusalem. Sewage from other Settler communities is regularly siphoned through Palestinian villages. Since 1967, the Israeli authorities have uprooted over 700,000 olive trees: about the same number of

Palestinians were uprooted from their homes in 1948. In damaging the land itself, Israeli settlers are in effect providing emotional evidence that they themselves are not treating it as their home.

For Palestinians, memory is necessary, while for the Israeli regime, it is better to have a sketchy memory for atrocities committed. It is probably quite hard to hold in mind the three-year-old Palestinian child who suffered a head wound and when, after two hours, doctors opened up the wound, smoke came out of her head. White phosphorous, in Gaza, 2009. Lest we forget. There was a similar pattern between the English and the Irish, as Éamon de Valera noted, saying that the difficulty was that 'the English never remember, the Irish never forget'. The vanquished are left with nothing but exile and memory while the victors are left with land which does not speak to them. Within Britain, the Irish, Welsh, Scottish and Cornish know a love of land which the English, so often, do not experience. Why so? In part, because the first acts of empire were internal, making Ireland, Scotland, Wales and Cornwall into colonies. But in the long run, it is the English who have become 'internal refugees' in terms of home-land-love.

Dispossessing, murdering and enslaving Indigenous people, and removing them from their lands, was the story of empire. But if one mentions the reckless cruelties imposed for the sake of empire, the chances are that someone will sneer that one is suffering from a hand-wringing post-imperialist guilt, as if guilt is an unhygienic bad habit, a perversion, a personality disorder. Should the British (and particularly the English) feel guilty for the atrocities of empire? Of course we bloody should. To abuse people without a flicker of guilt is something of which only psychopaths would be proud. After the empire, though, after a few guilty recollections, what then? We have, collectively, through neoliberalism, through corporations, through consumerism, through extractive industries and the arms trade, continued to crush the peoples of the world who we first impoverished by empire. Should the British feel guilty now? Of course we bloody should.

The psychopath insists that one should not dwell on the past. Onwards, onwards on the road to the future! But the past is where the deep truths of today were seeded. The past plays cause to today's consequence. Memory has become a political act and it is more radical to remember our history accurately than to don a balaclava and smash up a shop.

But for the English in particular, memory is difficult. We seem to want to remember Robin Hood, King Arthur and Puck, perhaps, but Olde England seems to be visible only as some cheesy Avalon seen through the windscreen of a BMW, or some beer mug with the Green Man leering in the handle. But the gods won't play. Something in the spirit of the land seems shy of us and will not grant us an authentic dwelling for the soul, as if some shame palls the land for us, as if our English indigeneity is something we want but cannot find.

You can do a straightforward Google search which gives one answer as to why the indigeneity of the English is a contorted feeling. When I first wrote about these issues, I typed in 'indigenous Britons' and seven out of the first ten websites were all related to the British National Party, the BNP. 'The liberal-left love to applaud Native Americans for their "soul and soil" approach to life, but we [BNP] reflect such an approach in our own Nationalist mindset,' read one. Blood and soil. So the territory of English indigeneity was stolen by the far-right. The human home-land-love is perverted into a hatred of other people's pigmentation and a queasy calculation of blood quota. For the record, many Indigenous societies have a history of welcoming those not related by blood, particularly those who want to live amongst them and according to their lifeways. (Indeed, the first word, spoken by the first Native American to make contact with the Pilgrims of Plymouth Colony, on 16 March 1621, was 'Welcome'. The second, third, fourth, fifth and sixth words were: Have you got any beer?)

UKIP, as nationalism's successors to the BNP, is less honest and more coded, asserting it is not racist even while openly encouraging xenophobia and, with a striking lie, claiming victim status, as if the English were the plucky underdog voting Brexit to escape a bullying European Union.

The 'nationalist mindset' and nation states are fake and recent political constructs, used frequently to attack, demean or destroy others. Land, on the other hand, is unarguable and unartificial. By dishonestly merging those two concepts, racism is poisoning, for all the English, one of the sweetest wellsprings of the human heart. Many living in poverty were encouraged to see immigrants as the cause of both hardship and a sense of alienation from one's own land. The cause of those has never been immigration, but rather massive wealth disparities, all forms of enclosure and a legalized system of land thefts, as Nick Hayes shows so brilliantly in *The Book of Trespass*.

What the British have done abroad, in the form of imperialism, has also happened within Britain and indeed to most of the English – the colonizing of common land by which the wealthy have made serfs of the rest of us.

Start with some contemporary facts of land ownership. 0.06 per cent of the population owns half of the countryside in England and Wales, as Guy Shrubsole reports in *Who Owns England?* We the Commoners are fundamentally homeless in our own land. That is as shocking as any other statistic on the apartheid of land rights. Further, the wealthy landowners have also propelled the factory-farming agribusiness which not only so damages insect life but strips the land of specialness so the hare, the 'stag of the stubble', becomes rare. Mountain hares are also killed because they carry a tick which can infect grouse. The enclosures, by which the vast majority of us were made internal exiles, were an outright theft of land from which we English Commoners have never recovered. 'Private: Keep Out' signs warn us off our own land, and even innocent ramblers find that landowners set dogs on them and order the walkers off the land, at gunpoint. I've experienced both when I've been walking on public footpaths.

The government aims to make trespass not just a civil but a criminal offence, increasing the already suffocating atmosphere for most people. The Traveller experience is worse, for their very way of life depends on being able to move around in this green and pleasant land without being criminalized for it.

If there are three things that make me cleanly proud to be British, they are the NHS, the BBC and the system of public footpaths. In lockdown, all three of them came into their own. I live in mid-Wales, and spend a lot of time mountain-biking on bridleways, walking or running on footpaths, searching Ordnance Survey maps for good places for wild swimming. I delight in this freedom at all times, but in lockdown I absolutely craved it. We all did. Rural dwellers like me had access to green space because this is simply where we live. The wealthy, though, seized access, illegally relocating to their second homes and causing a crisis for local medical services unable to cope with the influx.

The urban poor, meanwhile, had no such access and

suffered months of house-arrest, experiencing enclosure in a visceral way. Where were the commons, for the Commoners to walk, saunter, picnic and sunbathe? Long gone. The dementing claustrophobia could have been assuaged: open the golf courses, argued Guy Shrubsole and others. Open up the playing fields of the wealthy public schools. Open up every National Trust outdoor space. It didn't happen. When Johnson's government introduced the 'Rule of Six', by which no more than six people could meet outdoors or indoors, they made an exemption for grouse-shooting and hunting 'parties' costing £3,000 per killer per day. Rich and White, you may stalk the land.

Poor and Black, you may not. During lockdown, police in England and Wales were twice as likely to fine young BAME men under Covid-19 regulations as young White men for being outside. In non-Covid times, there is a strange Whiteness in that green and pleasant land. Just 1 per cent of National Park visitors are from BAME backgrounds, despite making up 10 per cent of the population. Rhiane Fatinikun, founder of the walking group Black Girls Hike, says Black people 'stand out' in rural areas, which can feel 'unwelcoming'. On one of their hikes, they were told to 'go back to the ghetto'. Poet Benjamin Zephaniah, meanwhile, went out for a jog while staying on a friend's farm in Essex. When he got back to the house, all hell had broken loose, the place was surrounded by the police, with a helicopter circling above. For why? Because, the police said, 'We have had reports of a suspicious jogger.' To be Black outdoors is to be suspicious.

'I wandered lonely as a Black face in a sea of white,' comments Ingrid Pollard in *Pastoral Interlude*, a brilliant photographic commentary on English landscape. The Black British experience is typically to be perceived as urban, part of the crowds in the built environment and therefore stereotypically denied the role of the individual in that posture of solitary reverie. It is not an innocent trope, for it displays the entitlement of wealth and the leisured classes, a sense of ownership: the gazer is lord of all he surveys, larger than life.

Actual statues adopt a similar stance, the gaze of dominance. The great historical era of statue-erecting is from the early eighteenth century, but public statuary enjoyed a golden age in the Victorian era, the height of empire. In Bristol, in the Black Lives Matter protests, the statue of Colston was pulled down and thrown into the harbour. In Richmond, Virginia, a statue of Christopher Columbus was toppled, set on fire and thrown into a lake. A sign was left in its place, reading: 'Columbus Represents Genocide'. Trump, who wishes to be immortalized on Mount Rushmore on land stolen from the Native Lakota Sioux, defended the statue of slaver Andrew Jackson, who forced Native Americans to the 'Trail of Tears' on which so many died. Trump claimed the anti-racist protesters were 'far-left fascists' wanting to 'wipe out our history'. In the UK, similarly, right-wingers claimed statues were necessary to teach history. So, Colston? Before the removal, most people in the UK would have been hard-pressed to identify him. Something to do with toothpaste? Or maybe mustard? But when the protesters brought him down, the public had an immediate and dynamic lesson: Edward Colston, slave-trader. This is how entire nations learn a bit of history, with unforgettable public panache and a *coup de théâtre*.

We all need to know our history. White people need to know Black history because there is simply no truth in White history without it.

For one long moment in my life, I heard the earthsongs of England. For one exquisite time, I saw the old gods honoured with an authenticity that left me in tears.

During the anti-roads protests of the nineties, the motley-wearers (artists, punks, shamans, squaddies, students, the homeless, pagans and peasants) fought for their land, literally putting their lives on the line when the authorities issued orders so reckless as to risk murdering the protesters. They wore the feathers of birds for the flight of the gods, they lit the fires of the solstices and paid raw tribute to the Earth. They picked up by ear the old songs, gentle as violets, tough as badger's teeth. Crucially, in every aspect of the protests they created a distinction – and an opposition – between the state and the land. They loved their land and hated their state, defied it with all they had, when the bulldozers came, building roads through the homeland of history, ripping apart the beauty which had graced those woodlands for generations. The protesters referred to the world of consumerism, cars and capitalism as the 'Babylon' of today.

To me, the protests were extraordinarily significant, rare as hares, signifying that authentic belonging to the land was something which had to be earned. To belong is to love, to defend with your life if need be. Among the camps – grubby, feral, crusty, sweet-hearted, pissed-up, kind, angry – the gods, who are never for sale, played. I'm no deist, but to me the gods were metaphor and symbol of the land's indigenous psyche. And they were green to the teeth, rampant and gurning.

One day, representatives of an Indigenous community from Bolivia, similarly campaigning against a road (the Pan-American Highway), visited one of the sites. 'We salute you,' said the Indigenous representatives to the protesters, 'as the Indigenous people of Britain. Your fight is the same as ours; you are fighting to defend the land.' It was honey to hear the term 'Indigenous Britons' used without racism.

There is a word so ancient that you can hardly say it in sneering metropolitan circles. It is *honour*. The protesters honoured the land which came alive in their honour. The bells woke the woodlands: wild horses rallied to the protesters. With their treehouses streaming ribbons, flowers and webs, the protesters created the Hanging Gardens in the Babylon of concrete and tarmac. From

Babylon, which, both ancient and modern, represents corruption, power and wealth, they wrested a corner for the mead moon, the Green Man of the woods, who here was neither toy nor relic but reignited, arising like a phoenix from the ashes of the Beltane fire.

It is as if there is a kind of Earth-ethic, an underground morality in this, whereby in order to experience one's home-land-love, in order for the indigenous human heart to belong profoundly to its land, there is a necessary sacrifice. It was exemplified, for me, by the sacrifices made by the protesters – the exposure, ill health, injuries, stress and burnout – but those are not the only people who know sacrifice to stay true to their lands. The hill-farmers of Wales, the crofters of Scotland, Cornish villagers and the smallholders of English farms (almost rubbed out by supermarkets and large landowners) also know the high price of fidelity to one's acre.

But those who steal the fat of other people's lands, who take far more than their fair share, and then ask for a taste of mead or the melody of the Ash Grove: those who gobble the resources of others and parade their lives of high consumption and then search for the god Lugh or the spirit of the woods, will find, simply, that they can't have both. You cannot take without giving, and if you refuse to give, then something will be taken from you, probably in the coin of the soul.

Indigenous cultures seem universally to recognize that every act has consequence. Where something is taken, something must be given in exchange. You cannot change the laws of physics, but you cannot change the laws of metaphysics either. For high-consuming lifestyles (which indirectly but certainly rob other people of their homeland) have a boomerang effect, causing a loss of belonging to the consumer.

I have tried to look at this in moral terms, but our language of morality (so good on person-to-person morality) seems to gloss over collective morality, and wholly omit an ethic of Earth. I am searching, then, to describe some truth of the psyche so profound that it is not only a psychological truth but also an ethic grounded in some irrefutable and intrinsic Earth sense.

For the English to have back our deep, lovely Englishness, we need to remember our past soberly, and to stop repeating its iniquities today through the devious reach of corporate colonialism. If we want to experience our home-land-love, we need to honour other people's homelands. We need to educate ourselves about our real history. We need to oppose our nation state for its racism, dishonesty and greed. We need to renounce the political and financial gains made from our nation's slave-trade, exploitations and wars. We need a situation where anyone can feel at home in the landscape, where no one feels out of place. Looking to the future, we need to make reparations to the countries which will pay a terrible price because of the carbon emissions we have created. If we want our English identity back, if we want to belong to our lands, we have to respect other people's. And at home, the wealthy, if they want to belong, have to return the lands to the Commoners. They won't, I know, but what they lose, without question, is the true belonging of the human heart.

While Indigenous people frequently say that they belong to the land, landowners claim that the land belongs to them. Ownership is an opposite of belonging and, for the large landowners of Britain, the more they own, the less they belong. Some of these large landowners evade tax by calling themselves 'non-domiciled'. Intended simply as a tax category, it in fact carries a far greater (and sadder) poetic truth. They are present-absent aliens.

For Indigenous Australians, 'ownership' of land involves not money but knowledge. The knowers of the land, the knowers of the songlines, are the true owners of it. Hedge by hedge, hare by hare, stanza by stanza and grove by grove, the land of England is there to be known, and there are those whose nostalgia for it hurts them. But as it was the very wealth of Babylon which both seduced and exiled Queen Amytis, so it is the wealth of a modern Babylon which seduces and exiles us all who, yearning for a sense of home, find that all the power, wealth and corruption of a consumer Babylon will not console this yearning unless some powerful restitution is made.

Let There be Dark – Soil and Stars

If I sleep out on the earth, I dream differently there. Soil touches my fingers, its damp sweetness rubs my back, its darkness beguiles my dreams. Soil is an unsung divinity, ubiquitous, utterly ordinary and yet the miraculous source of dark-dazzling life. Soil is where life is dreamed into being.

From the soil comes harvest on harvest. Sheaves of corn piled together reaching the skies; mountains of apricots; grapes a mile deep. This is the visible abundance of Earth's soil, but it holds subterranean beauty as well. The colours of soil include *oxisols*, the tawny shades most often seen in tropical rainforests; *spodosols*, the grey-yellow of boreal forests and heathlands; *ultisols*, that can be almost purplish, pale orange, low yellow or red; and *vertisols*, the earth of savannah or woodlands that may be grey, brown or deep, deep black, known as the black earths in Australia, or the black cotton soils in East Africa. Soil layers are called 'soil horizons': we look down to see the horizon. One soil horizon is called the 'eluviated horizon', a phrase to conjure with:

eluviated, rejuvenated, alleviated with the movement of rainfall, a transporting of matter further downward.

Underfoot, there are colonies of micro-organisms, countless and complex, in a lively world tingling with life. More than half of all terrestrial biodiversity is in the soil. Fungus (specifically mycelium) and worms, mites and bacteria create life's fertility and regeneration, and topsoil is created by creatures that can break down leaf litter and wood into the richness of humus, cycling food from deadness to make thriving life. From the dead back to the quick.

Who else dwells here below? Rotifers, their tails turning like wheels. There are protozoa, amoebae and nematodes or roundworms, some feeding on fungi and some being food for fungi and bacteria. (It's all feast down here.) There are 40,000 named species of mite. Here, too, is the champion of sheer survival, best known by its endearing moniker the water bear, and also the glorious Collembola, or springtail, that can jump a hundred times its own length. These miniature shapeshifting jesters can alter their size and shape rapidly if they need to – and they have eye patches. While some Collembola live on the top of Mount Everest, they are also the deepest land creatures on the planet, one species found in a cave in the Western Caucasus at a depth of 1,980 metres underground. They were coaxed out (reports Andy Murray, Collembola-lover, writer, macro-photographer, musician, juggler and archaeologist) with cheese.

The most visible and well known of the soil dwellers are the wriggly, glistening worms, making the soil soft, silky, supple and moist. Honouring their role in agriculture in the Nile Valley, Cleopatra proclaimed worms were sacred, and harming them was an offence punishable by death.

In Darwin's last work, *The Formation of Vegetable Mould, Through the Actions of Worms*, he sought to demonstrate that worms have mind. Worms make burrows, plugging the openings with leaves, and this activity, he saw, was neither random nor wholly instinctive, and in that space between, a form of mindedness can be inferred. Worms, he found, formed judgements about the different shapes of leaves, even the leaves of a plant with which they were unfamiliar: he witnessed the way they made decisions about how to best drag a leaf, by feeling it carefully and judging the right manoeuvre.

These unacknowledged makers of the earth, sensitive to touch like the tip of your tongue, are gardeners of delirious fecundity. They are the quiet ones, quietly making life possible, the ones who shun the spotlights, the unstarry ones on whose lovely liquid glistening – true brilliance – the world turns.

In the soil of my garden, under a rock, one of my cats is buried, the cat whose death left me inconsolable. Soil is not only where we come from but also the place where our existence ends, this dark forever where we go after death, earth to earth and ashes to ashes, our brief lives poised between two long eternities of soil.

I have watched eternity happening, rolling and unrolling time, turning it forwards. I have heard the little rattles of dead leaves, light as rain, as earthworms eat and cast leaf litter to create soil, acting in sweet complicity with both life and death. I have seen roundworms pass for shooting stars in shimmering slowness in the other dark beneath me. Life moves like this: quickening, creating, and turning. It is soil that turns the Earth's barren rock into the riotous life we know; wheels on wheels of life in this unique planet wheeling its green-blue feast through the starving blackness of space. Soil turns past into future, turns age into youth. *Viriditas*, the force of green life, grows out of the dark and turning earth, and worms are the artists of the turning world, returning vitality to it. Worm-shine licks the earth, tickling it to harvests and intricately re-thinking the soil into a different cast. The worm, passing through the soil and passing the soil through itself, turns soil into bright actuality, transforming the dark earth and turning death back to life, the lovely holy worm.

For Indigenous Australians, the soil contains the Ancestors, not only the human ones but the Ancestor spirits who travelled the land in the Dreaming, and the earth is alive with them still. Indigenous cultures very commonly assert that the land has consciousness. More: the land, they say, dreams us. This beautiful, enigmatic idea contains a vastness of vision, seeing our place on Earth as a kind of dream. The Dominant Culture is now rediscovering the knowledge that Indigenous cultures have long intuited: there is mindedness beneath us. It takes a certain cast of mind to find mind under the soles of our feet: it takes a kneeling humility to regard and to respect the mind beneath us, that has been called the 'Wood Wide Web'. Fungus is at the heart of it.

Fungus is a billion years old and it is through threads of fungi that a whole forest is connected and aware. This network connects almost all trees and plants in a web of information both gigantic and intimate, ancient and immediate. Tree roots spread out like curiosity, linking with mycorrhizal fungi in symbiotic relationship, feeding each other with information, messages from the underground, suggesting, hinting, telling, guiding, warning, giving, feeding the soil-mind with the oldest and kindest wisdom. This network of connections responds differently to the presence of different creatures.

Many of us have been wonderstruck at this discovery, for our minds recognize exploratory and communicative mindedness and, indeed, recognize its shape. We speak of message *threads*, communication *lines* and knowledge *paths*. We humans know that thought is well represented by linking filaments – including the tendril of this sentence through which my mind flickers briefly in yours – and in the soil the fungal threads make rhymes for our minds, underground.

The deeper we dig, the more life we find. Within the

subterranean world, over five kilometres deep, 2,734 fathoms down, there exists a rich ecosystem, twice the size of all the world's oceans, teeming with micro-organisms, hundreds of times the combined weight of all humans. The Earth has more life than previously thought. It is a discovery which takes my breath away. It alters my very footsteps: tread softly for we tread on the Earth's Dreaming, this subtle, ambiguous mystery surrounding all that we are, ancient and slow. Some organisms here can live for thousands of years, part of the unfathomably slow time of geology.

Modernity isn't good at slowness. It demands instant gratification and grabs 'right now'. It prefers shallow time to deep time and is impatient with the easy, languorous times of nature. All time lies below us in the soil: the deep past is there, the present is fed there and the future could be nourished there. Time is fed by soil.

In the slow time of nature, it takes about a thousand years for a new inch of topsoil to form (with no one watching). It also takes about a thousand years for a civilization to live and die (with no one really noticing). Why did so many unrelated civilizations all last about a thousand years? asks David R. Montgomery in *Dirt: The Erosion of Civilizations*. Because it took about that length of time to strip the topsoil.

When Rome was young, its soil produced a richness of olives, figs and grapes; peaches, apples and pears; almonds, walnuts and chestnuts. The land was the source of life both literally and conceptually, and the Romans called Earth our mother, *terra mater*, while many prominent Roman families took names of vegetables in earthy pride: the family name Fabius, for example, is from the Latin *faba*, bean.

But some philosophers proclaimed that the ambition of agriculture was to engineer 'a second world within the world of nature', in the words of Cicero, whose name comes from *cicer*, meaning 'chickpea'. Gradually, the soil of Rome was pressured to produce more and more. Over-ploughing led to erosion, and when there was heavy rainfall, soil was lost, slowly but certainly, more than an inch per century. Towns declined: populations left. 'Rome didn't so much collapse as it crumbled, wearing away as erosion sapped the productivity of its homeland,' writes Montgomery. In any one year, the amount of soil-loss would have been almost beneath notice: it took about a thousand years for the Roman heartland to lose its topsoil. 'We overcrowd the world. The elements can hardly support us. Our wants increase and our demands are keener, while Nature cannot bear us,' wrote Tertullian, in around 200 CE.

Across the world, from Egypt to ancient Greece, the story is the same: overuse, depletion, erosion and loss. When the great Mayan cities were built, people began farming the land to exhaustion. As soil erosion peaked, the Mayan civilization collapsed. Today, in the jungles of Mexico, Honduras and Guatemala, lie the tantalizing ruins eloquent of a lost culture, sleeping in the tangled green.

In northern China, forest land was stripped and cleared, and the topsoil was destroyed. It resulted in a massive famine that killed half a million people in 1920–21. The land was little more than dirt, barely dust in the wind, and some twenty million people were scraping the soil for anything at all that would grow. The inhabitants had starved themselves but 'just too slowly for them to notice', writes Montgomery.

'Upon this handful of soil our survival depends,' notes a Sanskrit text from about 1500 BCE: 'Husband it and it will grow our food, our fuel and our shelter and surround us with beauty. Abuse it and the soil will collapse and die, taking humanity with it.'

If I am lying in my garden at night, I feel a strong sense of two kinds of infinity. One is the infinity of outer space, stars and galaxies, calculated in light years and eternities that both dazzle and repel me. I cannot comprehend an eternity on that scale: human thinking is scaled to the niche of time that it can know and I prefer my eternities numbered not in light years but in Earth years. And I prefer the second kind of infinity: the infinity that sits snug in my hand, seeing, with William Blake, a world in a grain of sand and the infinity of the seed, the unending and unendable nature of life creating infinity from within itself, endlessly begetting oak trees from acorns, and happily replaying itself for ever.

If there are two kinds of infinity, there are also two kinds of darkness. One is the infinite lively darkness of the soil in which there is moist and glistening life, constantly regenerative. The other is the infinite blackness of outer space, the ultimate deadness. Of course, stargazing seems a universal delight, the human mind fascinated by the mystery of brilliant lights shimmering and distant in the infinite darkness above us. The starscape is a place of imagination, of the reaching, yearning tangent of the mind. *Further! Further!* Stars mesmerize my eyes. I am fascinated by the photography from the Hubble telescope. Looking at these deep-space objects, the billowing nebulas and entire galaxies that are so absolutely other, I falter, facing the unfathomable vastness: it feels like watching God with a kind of stricken awe, both aghast and amazed, and being soul-blinded by the enormity and the fearsome power of it.

The search for life in outer space obsesses modernity like never before. Of course the amazement of it has an easy appeal. Of course the 'miracle' of such a discovery would alter humanity's sense of its place in the universe, and the discovery would be an extraordinary tribute to the technical skills of the discoverers. And yet I loathe it because it feels directly related to how we are treating life on Earth.

The United Arab Emirates recently launched a spacecraft orbiter on a mission to gather data on the climate of Mars. And the climate of *Earth*? I want to ask them, is that of so little account? When water was found on Mars, the media went into a frenzy of delight. Meanwhile on Earth, a grandmother died of thirst in the desert in her homeland. Qoroxloo Duxee, an eighty-one-year-old, died of dehydration as part of the government's attempt to evict the Bushmen, as they call themselves, from their homeland in Botswana's Central Kalahari Game Reserve. Close to water but prevented from reaching it by the Botswana government's armed guards, she thirsted to death. The Bushmen are considered the very first peoples, the ancestors of us all, the grandmothers of us all, and yet they and other Indigenous peoples are routinely abused and subjected to genocidal policies precisely because of their attachment to the soil. Notably, land that is managed by Indigenous peoples is in general declining less rapidly than other lands.

The obsession with looking for life in outer space is precisely related to the killing of life right under our feet. We are being given baubles of distraction, so that our attention is drawn up and away, and so that we ignore human-induced mass extinctions, climate change and the death of the soil and the consequent threat to the food supply, here, now. Space research glitters and shines, alluring the public mind with a lurid and fatuous light show, a *son et lumière:* a spectacle of brilliance, and a spectre of the future. You can't eat light, no matter how brightly it shines.

*

The water bear is found in almost every habitat on Earth, from Himalayan mountain peaks to sea floors, from tropical rainforests to the Antarctic. It is also the first known animal to survive in outer space, on the outside of a space rocket. Water bears, so-called for their barrelling rolling gait, are more properly known as tardigrades, literally 'slow-steppers': not for them the speed of a rocket launch. Slow and ancient, they are thought to be some 530 million years old. About half a millimetre in length, they are chubby, with eight legs, and many have pigment-cup eyes and sensory bristles. They can survive cold at minus 272 degrees Celsius and heat at over 150 degrees Celsius. They can go ten years without water and thirty years without food. They can withstand pressure up to 1,200 times atmospheric pressure and have survived Earth's first five mass extinctions.

It seems like a parable. Yes, the water bears survived exposure to the vacuum of outer space without the protection of atmosphere, but they did so by suspending their metabolism, entering 'cryptobiosis': their own death-zone. As soon as they arrived back on Earth, they rehydrated in delirious relief with the water of life, as eager as every scrap of life is to thrive and flourish in the one place where it can. Imagine if just one such creature were discovered already dwelling in space: it would be treated as a tiny superstar. But because it exists already, here below us and in multitudes, as lowly as it is necessary, it was literally overlooked until it succeeded in surviving in outer space.

Modernity is not only obsessed by searching for life in outer space, but by attempting to create space colonies for humans to live in. Elon Musk writes of his plans to establish a city on Mars by 2050, saying: 'It's going to be hard . . . It's a very harsh environment so there's a good chance you die there. We think you can come back but we're not sure.' These words presage death, but not on Mars. For us all, on Earth, it is going to be a hard road ahead, with extinctions and climate chaos. There is a good chance of death. By that same year, 2050, if humans have not acted with a sense of emergency now, large swathes of the Earth will begin to be uninhabitable. We think we can come back from the brink of this, but we're not sure.

Elon Musk acts like a cipher for this age, seeking a crazed techno-salvation in the face of the unhallowed horror that is climate chaos. In this context of crisis, space travel is a morbid voyaging, a deadly rejection of real life on Earth. Nothing was more richly, vitally alive than Earth: nothing deader than space: it is death-post-death, the death from which no life comes, the utter and final black.

While vast amounts of money, energy and prestige are squandered in the search for a glimmer of potential life out in the deadness of space, or the search for a planet that may be capable of supporting human life, we are killing the very source of vitality, the proliferating life in the precious soil. The possibility of life on Venus was excitedly announced at the same time as a new study, the Living Planet Report, 2020, showed how more species of life on Earth were going extinct. In further bitter irony, Venus was temperate for two billion years, but it was turned into a hell with clouds of 90 per cent sulphur, and heat above 450 degrees Celsius. What changed? Run-away climate change. Far from finding life in outer space, we are spreading the deadness of space across the Earth. The fascination with stars seems intricately related to contempt for soil.

Only sixty harvests are left before soils are too degraded to feed the projected world population, the United Nations has warned. The forecasts vary: some say we have a hundred harvests and some say thirty. Inherent to the idea of harvests is that even if they fluctuate, they will always happen. Harvest means abundance in time, as perennial and certain as autumn, eternity on Earth. It is grotesque to imagine this limitlessness made finite.

Soil is abused, poisoned with chemical fertilizers, fungicides, herbicides and pesticides, leading to a collapse of soil health, the killing of microbiota. Intensive farming dims the vibrancy of the underground world; soil loses its organic components. Deforestation leads to soil erosion, and ploughing can damage the structure of soil, while the sheer weight of giant tractors compacts the soil and crushes the tiny dwellers within it.

Among them are the earthworms who increase

food-crop yield by an average 25 per cent and are known to be essential for soil fertility. An average field may contain two million earthworms and their abundance indicates the health of the soil. They eat and cast, eat and cast, and their casts are beautiful and moist, shining like diligence. But within three weeks of glyphosate application, their casting activity nearly disappears, as a study from the University of Natural Resources and Life Sciences in Vienna has shown. Their reproductive activity falls by more than half. In the overuse of herbicides, humanity is acting against its own interests and also against its own knowledge.

Soil contains life and creates it, not only directly with agriculture but in the larger realm of the global climate crisis, for soil holds more carbon than all of Earth's plant life, and it can absorb more, depending on the life within it. In devastated soils or with the excessive use of synthetic fertilizers, the balance switches from absorbing carbon to releasing it. The relationship between carbon and soil is so entwined that the climate crisis cannot be halted while the soil continues to be devastated. It is that simple and that vital.

About a third of the world's soil – approximately the surface area of Mexico and the USA – is thought to have been degraded over the last forty years. When soil is degraded, it is more likely to suffer erosion from wind and rain. While Europe loses nine million metric tons of soil each year, the World Wildlife Fund says that half of the world's topsoil has been lost in the last 150 years. Because soil is created so slowly, it is in effect a finite resource.

We know what happens. The Great Plains in the United States were once famous for their fertile prairie soil, but the area was first over-grazed and then over-ploughed. In a drought lasting eight years in the 1930s, the soil simply had no strength left. Exhausted and depleted, it couldn't hold itself together and blew away to nothing but dust, the dust bowl of a hundred million acres that lost all or most of the topsoil. 'Forests precede civilizations, and deserts follow them,' said Chateaubriand.

The information is available. The facts are there, the lessons from the past that warn about the future. And yet all this knowledge is not enough, because collective attitudes run counter to them. The Dominant Culture's mind-set remains colonially entitled and extractive, seeing soil as no more than resource. Interestingly, in the famine of northern China in 1920–21, when a foot of topsoil had been lost from hundreds of millions of acres of land, there were exceptions: around the Buddhist temples, the soil was deep, black and rich in humus, because the Buddhist philosophy stipulated the protection of the forests.

You are what you eat is true in body but also in mind, for you become what you think, and what you attend to, what you tend, tends to flourish. If you attend the feast of the soil then you are feeding your mind with a green thought in a green shade, in Marvell's words. If, though, you feed your thoughts with mental glyphosate then your mind will operate a toxic sterility. Attend the famishment of space fantasy and the mind mimics the superficial artifice, a Teflon thought in a metallic glare, gazing up into a starving sky.

Indigenous cultures, typically, cherish the Below as a source of life, wisdom, dream and ethics: 'Law comes up from the Land,' Indigenous Australians say: what is underneath us teaches respect for all life, but the Dominant Culture privileges what is above rather than what is below: it looks up rather than down. We pay respects to Her Royal *Highness*; we pray to God in the *highest*, and aim for the *heights* of success. We speak with contempt of the *lowest of the low*, and Donald Trump refers to Black Lives Matter protesters as *low life*. So soil is treated with contempt because it is 'beneath us'. We English speakers misuse the word 'soil' so that something 'soiled' expresses not just disdain but visceral disgust, speaking with revulsion of the very thing that sustains us.

While heaven is above, hell is below. The Fall was our tumble out of an unearthly paradise on to mere earth. Although Adam is named from the Hebrew 'earth', as if the wisdom inherent in language was compelled to acknowledge our identity with soil, yet Genesis itself revolts against earth; the lowest of creatures is the worm, creeping on the ground.

'Never use the words higher or lower,' Charles Darwin wrote in a note to himself in the margin of a book he was reading. 'It is absurd to talk of one animal being higher than another,' he wrote in his notebook. Writing in praise of worms, he said: 'It may be doubted whether there are many other animals which have played so important a part in the history of the world, as have these lowly creatures.' His courage took many forms. Not only in his theory of evolution, but his lifelong love of the worm. 'When we behold a wide, turf-covered expanse, we should remember that its smoothness, on which so much of its beauty depends, is mainly due to all the inequalities having been slowly levelled by worms,' he wrote.

Levelling inequalities conjures a social and political rhyme, suggesting the level earth as a level playing field, and the levelling of unequal lives. The spirit of levelling holds humility close to its heart, it has social grace and ancient wisdom. Worms, the original levellers, are creators of worlds. And we humans are destroyers of the only world we have.

If I sleep out in my garden, I am aware of the privilege of having one, but access to earth is something we all need for we, like trees, need our roots in soil. Our minds are influenced by the scent of earth – smelling soil makes us happy, as the microbe *M. vaccae* increases serotonin levels. To be human is to need earth.

Indigenous people often express their Earth-ethos in that simple but deep observation 'we are the land', speaking as 'people of the earth' who retain their humble relationship to the humus, the earth, that commonwealth of humility under our feet. The words 'humus', 'human' and 'humility' are related: cognate words. 'Humus' means 'earth' or 'ground'; 'humble' means 'lowly' because it is near to the ground; and 'human' reminds us that we are made of earth. If humility were imagined as a visible thing, it would be soil: quiet, brown, soft, maintaining networks underground and feeding the whole of the living world.

This is where tranquillity lies and where we can find the sweetest sleep, and a humble dream. Sleeping out, especially by water, my dreams are bigger than me, my ego-life is just a little leaf litter on the surface, while the deep soil gives me mythic dreams. I am not alone in this. There is something about the darkness of the earth that touches the spirit.

In ancient Greek thought, when anyone falls asleep or dies, their memories are surrendered to the waters of Lethe in the Underworld, the waters of forgetting. These waters and all the memories they contain would slowly sink into the earth, pass through it, drunk by the earth until they come up again in the waters of a memory: the memories have been remembered, rethought, by the earth. These thinking waters, remembering waters, were called the waters of Mnemosyne, who then gave birth to all the muses. All memories, held jointly, remembered by the earth, then inspire the thoughts of worlds to come.

Each dreamer is connected to a subconscious wisdom extending far beyond themselves, like roots extending into a communal shared intelligence under the soil.

Insufficient sleep leads to cognitive impairment, it damages short- and long-term memory and diminishes our capacity for empathy. 'When people do not dream well, in a very fundamental way, they stop growing,' writes Rubin Naiman of the Center for Integrative Medicine at the University of Arizona. Dream-loss, he says, is 'an erosion of consciousness'. Meanwhile 'dream eyes transcend waking egoic perspectives, opening us to greater social and spiritual consciousness'.

Sleeping is the subsoil of the mind, the darkly luminous place for insight beyond obvious sight. Dreams, like worms, process the past, and create the future: dreaming means sifting through fresh material, turning things over, mulling things over in the humus of the mind, recasting things to see their deeper truths in the numinous world below. What soil does for physical life, dreaming does for the mind, and dreams are another kind of food in another kind of soil, the dreamosphere in the psyche, the deep dark of the sleep-soil, from whose depths both goodness and wisdom can grow. The land dreams us, truly. We who live on Earth, whose minds are soil-wise by starlight, are both the dreamers and the dreamed.

A Time of Rebellion

With a pocketful of superglue, anything is possible. I am sitting with my partner in the middle of the road outside Parliament, facing a line of police, the full moon shining over Westminster Abbey. A sound-system in a trolley is being towed around Parliament Square for the amusement of rebels. The police have just put in a request to the DJ to play 'Thriller' and there is a sinuous wave of dancing around a grandmother in a wheelchair, who is here because she is trying to get arrested.

Superglue is good for bonding, superb for attachment. My partner and I have a bonding moment, courtesy of a few squirts from the tube, and we hold hands tightly, stuck fast. 'We're literally handfasted,' I laugh. For me, it is the second-sweetest moment of this sweet rebellion.

We are part of Extinction Rebellion, aiming to block the streets of central London. We are rebels, gluing ourselves to almost anything so it is harder for the police to remove us. In the warm, streaming sun of early summer, people are glued to the headquarters of an oil company, glued to the Stock Exchange, glued to the tarmac of the streets. Like Antaeus with superglue, we gain our strength from ceaseless contact with this good earth. Love and holding fast to the ground we stand on, laying our bodies down as roadblocks, using anything – a boat, an open-sided lorry – as an impromptu stage in the middle of the streets, the rebellion is designed to block business as usual, because what passes for normal is not normal and cannot continue: the very laws of physics preclude it. So while the juggernaut of our current way of life is bearing down on us, Extinction Rebellion (XR for short) will stand in its way.

The Houses of Parliament are shrouded in grey veils. This is for practical reasons, for repair work, but it has a symbolic aptness: this occluded ghost speaks of a political system that is moribund. Rebels are using the shroud as a screen to project David Attenborough's film *Our Planet*. As my partner and I hold hands – firmly – the film shows a father and son driving through wildfires in California, the vehicle seemingly about to explode with heat, the flames leaping at them. The film's message is ours: we are taking a road that leads to hell; we have to stop and turn back.

On the first day of the rebellion, some 10,000 rebels, nervous and expectant, gather on pavement and street corners, at various prearranged sites from Oxford Circus to Waterloo Bridge, Marble Arch to Parliament

Square. At a minute before the symbolic eleventh hour, a man passes softly through the crowd. 'In two minutes,' he whispers, 'be ready to go.' As the pedestrian lights turn green, we surge on to the streets and simply stay there and, doing so, slide a new world into being. The police, who have been openly informed of our plans, step forward to talk to the drivers, beautifully conveying our message: Stop. Turn back. The way ahead is closed. And closed it will remain for ten days.

The streets become a surreal theatre. Central London streams with XR flags in luminous yellow, green, blue and pink. Pavements, walls, and clothes are branded with the instantly iconic XR logo. One thing is incontrovertible: this rebellion has style. The symbol is the circle of the planet containing an hourglass, the sands of time running out. It is edgy and exact, flexible, easily reproducible, identifiable and simple – a cultural meme of pure genius. The fluorescent colours have a challenger quality right up to the edge of provocation. Skulls are used as well as bees and butterflies. XR's style is psychedelic, invigorating, intoxicating, carrying an electric message: this is an emergency.

At eleven minutes after eleven, a bright pink boat of pure panache is towed to the centre of Oxford Circus, and a cheer goes up as its mast is raised. Why the boats? Rising sea levels mean that in the future boats may well float across central London. More, XR, Noahesque, stands as an ark for the more-than-human world and its leaflets and public information offer lifeboats of ideas for the real emergency we are in. *This Is Not a Drill* is the witty title of the XR handbook. *Ceci n'est pas une pipe.*

XR makes the future immediate and now. How do you raise the alarm? By showing that you are alarmed. Any self-respecting meerkat knows this. No sensible mammal would use an impersonal complicated data set to tell its fellows they are in terrifying danger of slaughter. Climate scientists thought their studies would be enough, that action would be taken. The scientists had no choice but to stick to the values and register of scientific communications, but too many people in the wider world adopted not just the facts of science but its demeanour and tone: level, unemotional, flat-factual. The opposite was required. The facts, yes, but not the voice. What was needed was an outcry of alarm. XR stepped up.

On Waterloo Bridge, full of woven willow and flowers, cyclists and a pop-up kitchen in blazing noonday sun, I speak to academic Rupert Read, a key figure behind XR's thinking. 'The most important thing is not to trust too much to hope, at this point. For thirty years, campaigners have gone for passive wishing and have been set on not scaring people,' he says. The result is mass complacency. Hope paralyses. Alarm galvanizes. 'We need something more important than hope now, and that is the courage to look at this brutal reality of the situation we are in, and the worse reality that is coming.' Indeed, a meta-analysis of 152 studies in public health (by communications experts Kim Witte and Mike Allen) shows that fear changes attitudes. It is also emotionally congruent to the facts.

For the facts are terrifying. Raise the alarm for what the Smithsonian has called death 'on an unimaginable scale'. At a two-degree warming, island nations will be victims of a genocide. Kevin Anderson, director of the Tyndall Centre for Climate Change Research, says: 'For humanity, it's a matter of life or death . . . If you have got a population of 9 billion by 2050, and you hit a rise of 4.5 or 6 degrees Celsius, you might have half a billion people surviving.'

Language buckles in this heat. What is happening is unhallowed. It is 'a drama unfolding at theological scale', says Roger Hallam, one of XR's co-founders. The governments of the world are taking us to our deaths and, say many in XR, there can be no greater crime. The grief of it all overcomes me sometimes. One half of all species could be forced to extinction by 2050. Sometimes I feel that I am in a constant state of farewell to the living world, knowing that it will not fare well at all.

XR launched on 31 October 2018. The date was chosen for its resonance with Halloween, All Souls and the Day of the Dead to underline the lethal situation we are in. October 2018 saw the publication of the IPCC

(Intergovernmental Panel on Climate Change) report stating we had twelve years to turn things round, presenting, says Hallam, a 'new realization of immediacy. That was the crack in the collective psyche.' Twelve years is also the lifetime of a young person.

The spring rebellion was chosen to coincide with the full moon of Easter and Passover and with Earth Day. April is the time to plant seeds, the season of growth, and the young are at the heart of it.

'Imagine,' says Robin Ellis-Cockcroft, twenty-five, who organized XR Youth, 'the grief of knowing that your grandchild might not exist.' Fellow XR Youth rebel Savannah Lovelock, nineteen, says: 'The baby boomers literally stole our future to create their present. They stole our dreams. A lot of hurt comes from that. Adults don't know how lucky they are to have lived their lives looking at the future and *not* seeing death right there.' Young people hold a banner that reads 'Are we the last generation?' and a twenty-one-year-old says to me bluntly: 'Young people are angry with older people. You fucked this up for us.'

The young speak with defiance, at once powerful and vulnerable, and with clear-eyed courage and unimpeachable moral authority. None is more famous than Greta Thunberg, who arrives during the rebellion to speak to the crowds. Millions of young people now follow her in the Youth Strike for Climate movement, marching on to the streets and demanding their primal human right: to live. Some are going on 'birth strike', too frightened to even consider having children.

During the rebellion, crowds of children sleep in tents at Marble Arch, making this busy traffic junction a gentle but enormous impromptu campsite. It is awash with XR flags, and I have Leonard Cohen's 'Hallelujah' in my mind: 'I've seen your flag on the marble arch.' On Good Friday, while arrests are being made, the children march to Oxford Circus as moral support.

Honouring the protest movements that preceded it, XR notes how the youth climate movement is this generation's version of the Children's March when, in 1963, children and young people of Birmingham, Alabama, poured out of school and on to the streets, willing to speak their truth to power and claim their rights. Martin Luther King is frequently quoted. Black Lives Matter is referenced. Like Gandhi, XR believes that civil disobedience is a 'sacred duty', while it learns from the anti-capitalist movement and from Occupy to stress the relationship of the 99 per cent to the 1 per cent. XR has the street-smarts of Adbusters as well as their *détournement* of the clever reversal. There is a lot of EarthFirst! courage and commitment here, as well as Greenpeace in its 1970s guise. It takes from Climate Camp the importance of using regional gatherings and from the anti-roads protests the physical defence of nature, as those '90s protesters used treehouses and aerial walkways to block bulldozers. From Reclaim the

Streets, it learns festive blockades, with pop-up kitchens and skateboard ramps and playgrounds in the middle of the streets. Many protest movements have been damaged by an overuse of alcohol and drugs, leading to sometimes chaotic atmospheres. By contrast, XR bans alcohol and drugs on site so people really get their shit together. Literally. At Oxford Circus, compost toilets were constructed within minutes, and if there is one single tribute to XR's organizational skills, it is this: in ten days of rebellion, the toilet paper never ran out.

A provocative edge of punk cuts through XR in its fluorescent fuck-thissery and its utter refusal to play by the conventional rulebook. It also has a strong scent of patchouli – openly delighting in its hippy roots, steeped in herbs and flower power – and takes a leaf from the book of John Lennon on the Establishment: 'The only thing they don't know how to handle is non-violence and humour.'

There is a red thread that runs from the women's peace camp at Greenham Common, to XR Peace, campaigning against militarization. XR looks to the suffragettes not only for its gender politics but also for its illustration that society is changed through direct action, not through asking nicely. More widely, XR pays tribute to the primal female, the divinity of the Earth as Mother.

On the first day of the rebellion, the cathedral of Notre-Dame catches fire. It is a poignant symbol of the burning of cultural beauty, for with climate collapse comes civilizational collapse. It is grievous and uncannily apt: Notre-Dame means 'Our Lady' or 'Our Mother', and as the cathedral burns so the Mothering Earth is burning: the verdant mysteries, wisdoms and medicines of the Amazon going up in flames, wildfires out of control in the Arctic. One of these mothers can be rebuilt with money, the other only with a radical change of heart, and XR dedicates itself to this, like a chivalric knight-errant to his lady.

When I first came across XR, I caught the scent of something both ancient and strikingly new: chivalry. The building of Notre-Dame commenced in 1160, the same decade when the code of chivalry was created. This code includes courage, honour, devotion, self-sacrifice and a willingness to endure hardship, and XR follows it while serving Our Lady the Earth. I hear this rare but unmistakable tone of chivalry everywhere. Rebels frequently say they are 'in service', talk with humility in the face of transcendence. Ronan McNern, who brilliantly leads the media team, says: 'The work I do is an honour. It's about service.' Many speak of the importance of virtue-ethics, doing the right thing, no matter the outcome.

That night, as Notre-Dame burns, it is freezing cold at Oxford Circus. A group of us are on night watch. It's three in the morning, and a friend and I are cuddled together for warmth in a hammock slung between two traffic lights at the top end of Regent Street, laughing in sheer amazement at the audacity of the Pink Boat. To protect it from being towed away by the police, rebels have superglued themselves to its underside and are immediately known as 'the Barnacles'. With ringing simplicity, the words TELL THE TRUTH are stencilled on the side.

'Tell the truth, and act as if that truth is real' is XR's first demand, asking that we truly inhabit the horror of our situation. They quote Gandhi's 'truth-force', Satyagraha, firmly holding to truth in words and respecting the importance of what is real and good: 'the Force', in Gandhi's words, 'which is born of Truth and Love or non-violence'. Truth is central to ordinary, decent human values. It is taught to children, and encoded in spiritual law and mythic lore – 'the truth will set you free' – and yet, in this age of widespread advertising deceit, where the current UK Prime Minister is an avid liar and Donald Trump uses lies as a malignant strategy, and where climate scientists receive death threats for their fealty to truth, in this post-truth world, where the media routinely distorts or ignores both the climate crisis and extinctions, truth is now a rare thing in public life, as radical and unexpected as a bright pink boat anchored in Oxford Circus.

The boat is deliberately placed within sight of the BBC's London headquarters, and XR's campaign, titled 'Media, Tell the Truth', includes a 'Requiem for a

Dead Planet' outside the offices of the *Daily Mail*, and also the *Metro*, *Evening Standard* and *Independent*. With skeletons, eulogies and speeches about those already dying because of the climate crisis (about 315,000 each year, according to the United Nations estimate) the requiem uses a live baroque music performance, and places lilies everywhere. While the media as a whole is fascinated by intellectually feckless and factless info-tainment, XR speaks otherwise. There can be no bigger news story than the climate crisis, the potential collapse of civilizations, the extinctions of species, and the threat to humanity's existence. The media attack us for being 'too middle class'. This is code. What it means is that we are educated and self-educated, proudly so. The media's coded attack is part of its intellectually thuggish attitude to the expertise and science of climate.

Ribbons of scarlet at their wrists, the Red Rebels sweep across Piccadilly Circus. They are street-theatre performers dressed in red with chalk-white faces, who flow through the crowds then take up positions, forming *tableaux vivants*, living sculptures. Each performer wears a headdress, a chaplet of red roses, long gloves and veils. Red is symbolic of the spilled blood of children and other species. Red is a warning sign of danger ahead. *Stop. Turn back. You are going the wrong way.*

The Red Rebels perform in silence, and through silence they speak. Scarlet lips and black eye paint against the white exaggerates facial expressions as they move slowly from one emotion to another, using bodily gestures of grief, love, justice, fear, joy, pity or victory. Collectively, they distil that feeling into an intensity and transcendence that one person alone cannot create. The sombre effect shudders and astonishes the crowds that gather around them. Powerful, ambiguous, archetypal, the performance uses aspects of Japanese Butoh theatre and recalls the Furies and the Divinities, the mythic psyche on red alert, burning and bleeding. Like the Chorus of a Greek tragedy that represents the ancestors or the unborn, the Red Rebels assume the conscience of the past and the future, seers whose gaze reaches to the far horizons of vision.

First I watched them, then I joined them. In costume, and only in costume, I felt within myself how profound an experience it is to step out of time, speaking from a place far beneath, and a time far beyond, this moment, how powerful it feels to lose one's individual voice, courting silence together to create that other-worldly presence.

Meanwhile, back in the now of the rebellion, the police are arresting eighty-three-year-old grandfather Phil Kingston. He is part of Christian Climate Action and one of many rebels with a strong faith. Supporters of XR include Buddhists and Quakers and the ex-Archbishop Rowan Williams. The radical rabbi Jeffrey Newman is arrested for XR in his rabbinical robes, kneeling and praying, and imams and vicars have joined the organization's ranks. XR's vision includes a core of Sufi thinking: unflinching truth, reckless beauty and audacious love. Roger Hallam often looks like a flame-eyed John the Baptist, albeit one who frequently interrupts himself with a huge and self-deprecating laugh. He uses the language of religion and refers to the nineteenth-century 'Great Awakening' for this 'woke' generation: conjuring a sense of religious revivalism. XR has an indubitable messianic quality; indeed, how could the vital need to save humanity and the natural world *not* have a salvationary note?

On the third day of Rebellion, I watch from the crowds at Oxford Circus, as Daiara Tukano, resplendent in beads and feathers, takes the microphone at the Pink Boat. She is from the Tukano people of the Brazilian Amazon, now so cruelly threatened by Bolsonaro. She speaks of the earliest Earth protectors: Indigenous peoples. It is they who protect 82 per cent of the world's biodiversity, they who have led the resistance to climate chaos, they who are disproportionately affected by it, and they who are in grave danger for their protest. More than 1,700 environmental activists have been murdered this century, a disproportionate number being Indigenous people. The Pink Boat is named *Berta Cáceres* after the murdered Honduran Indigenous activist killed in 2016 for fighting for the land and water rights of the Lenca people in Honduras, fighting for their home.

Hindou Oumarou Ibrahim is an Indigenous Chad

pastoralist. 'I will likely be part of the last generation of Indigenous peoples,' she says. Interviewed, she speaks carefully and with exactness. Almost imperceptibly, her eyes well up with tears. 'The third world war is the environmental one and is much more criminal than the first and second because the third extincts not only peoples but all the environment.'

Imperialism and colonialism have caused the genocide of Indigenous peoples for the sake of extractive industries. In terms of the climate crisis, the wealthy nations have sown the winds and are forcing the poorest to reap the whirlwind. One XR banner reads: STILL FIGHTING CO2ONIALISM: YOUR CLIMATE PROFITS KILL, for XR consciously seeks to educate both the public and indeed itself about the politics of the climate crisis, committed to rebelling with cause. Rebelling with compassion. Rebelling with creativity.

Its rebellion is infectious. XR has grown to over 100,000 members in the UK and spread internationally to fifty-five countries. In the US, rebels protest against *The New York Times* and, in Washington, superglue themselves to doorways at Capitol Hill, blocking members of Congress attending a vote, seeking to direct their attention to a motion of climate emergency. What other choice do we have when our politicians choose money over life? they ask.

XR canoeists in Australia are taking to the waters to draw attention to the Great Barrier Reef, and giant XR symbols are springing up around the world: a sand sculpture in New Zealand, an XR symbol made of flowers in Medellín, in Colombia, and in New York one made of human bodies. In Zurich, XR rebels dye the Limmat river a luminous green to draw attention to the impact of climate change on global water supplies. Across the world, people are willing to be arrested for their protest.

I spend days trying to get arrested, along with over 1,300 others, including a marine biologist who is seven months pregnant, carrying her own ocean inside her. On Easter Saturday I am lying in the middle of Oxford Circus with my arm inside a 'lock-on pipe', and I'm here for the duration. This pipe is metal, dipped in concrete, wrapped in roofing felt. Inside is a bar, so if you have a chain around your wrist and a karabiner, you can lock yourself into the pipe, and to free you the police will have to cut your arm out of the construction. It is a sophisticated kind of superglue.

Around us, children are writing notes to the arrestables. One reads: 'I don't know you, but I love you for doing this.' The noon sun is hot: so hot that people are putting up umbrellas as parasols to cover the heads of those locked-on.

The sparks from the police angle grinder fly around my hair, and the heat is powerful. I am suddenly frightened that they could misjudge the blade and my right hand, my writing hand, could be damaged. One police officer regards my situation with bewilderment. 'You could just release yourself,' he points out, reasonably. 'But then I probably wouldn't get arrested,' I say. 'Why do you *want* to be arrested?' It's an important question.

The willingness to be imprisoned for non-violent civil disobedience is one of the most powerful ways to bring about change. 'I've prototyped prison as a campaign strategy,' says Hallam. If the police's strength is their ability to arrest, XR's strength is to say 'yes, please do' and immediately, by seeking it, to co-opt the strength of the police. Non-violence is crucial: as research shows, 54 per cent of non-violent uprisings achieve their objectives while only 25 per cent of violent ones do. Arrest is, says Hallam, 'the classic sacrificial move'. Once the public see large numbers of people taking the climate crisis so seriously that they would give up their liberty for it, the severity of the situation is underlined.

Not everyone in XR is courting arrest: it is a sad truth that people of colour have good reason to be far more cautious of any involvement with the police and justice system than Whites. I am White and middle class, and the fact that the police are likely to treat me better is an uncomfortable privilege, but the tactic itself works because the media pay attention, covering the climate issue as never before. Hallam adds: 'In the UK, when we take to the streets, we write a solicitor's number on our arms. In other countries people write their blood type. Others just write their names. Who are we not to act when we have the freedom to do so?'

In the hour it takes for the police to cut through the lock-on, I ask an officer to get a message to my partner to sing me something. He, knowing the soundtrack of my mind, cups his hands and begins: 'I've seen your flag on the marble arch / And love is not a victory march / It's a cold and it's a broken Hallelujah.' A policewoman is kneeling by me, asking if I am okay. She has tears in her eyes. 'He's got a beautiful voice,' she says. I nod. Many in the crowd are in tears, and some sing with him, until hallelujahs ring out beyond us, beyond the police, out into the world. *This* is my sweetest moment.

Cell number 12.

The door clangs shut. Silence. Solitude. I feel like crying.

I talk to the police when I can. One says this week has been 'rejuvenating' for him, and another says: 'The thing is, you're all so *nice*.' This is at the heart of it. XR is committed to a simple strategy: be nice to the police. They see it immediately and that, as well as being strictly non-violent, organized, and tenacious, is a novel thing for the police. On the first night of the rebellion one officer had summed up for me how he saw it: 'Your attitude is good, so our attitude is good, so your behaviour is good, so our behaviour is good.' But this is more than strategy, it is the truth: we *are* all in the same boat together.

The cell is square and tiled, with a toilet in one corner and a sleeping shelf along one side with a thin blue plastic mat and a blue blanket. The door has a three-inch-diameter peep-hole, so officers can look in. There is also a rectangular letterbox through which they post food, water, blankets. There is a strip of fluorescent light. Extinction is like a prison cell, the bleak and lifeless tiled veneer where nothing lives. Not a shred of earth, except a little potted plant that one of the rebels had with him when he was arrested. The police look after it for him.

The officer doing my ID records (I've never been arrested before) is kind and gentle as he takes mouth swabs and fingerprints. Another asks me about my wool-plaited wristband. A little girl gave it to me to say thank you. 'Ah,' he smiles warmly, 'there is a God.'

Time goes strange in a police cell. I lose all sense of it. It's hard not to be in control of your own hours, not knowing how long this will last. I dread being released at three in the morning. It is hard on the psyche, no question.

I sing. Leonard Cohen is with me. 'Like a bird on the wire, like a drunk in a midnight choir, I have tried in my way to be free.'

I write.

I sleep.

Then dawn breaks on Easter Sunday morning; all the light of the earliest intimation of summer is pouring across London in the sunrise. This light, this time of year, streams with hope and possibility: the sun itself streaming a ceaseless and potent *yes*, and I am released from police custody at 5 a.m.

The police desk sergeant refers to 'your fellow-protester', and I correct him, saying, 'Fellow *rebel* – this is a rebellion.' And he says, 'Sorry, yes.' He looks a little awkward, embarrassed, as one does when using an unfamiliar – taboo – word, one that is also 'sexy and transgressive', according to Hallam, which is why it was chosen. As we leave, the rebel is reunited with his little green seedling, and one of the officers smiles candidly: 'You lot have been the best bit of my job in all my life. We're on your side.' An officer shakes my hand as I leave. 'God bless you. Good luck.'

I walk to the tent village at Marble Arch, where a rebel is fast asleep in a treehouse and, without the noise of traffic, I hear the birds singing a dawn chorus of exuberant joy, birdsong here for the first time in decades. It is miraculous to hear after the cold silence of the cells. When the rebellion ends, the campsite is tidied up, so nothing remains. Nothing, that is, except something potentially priceless: an unconfirmed Banksy artwork that miraculously appeared overnight. In it, a little girl holds the XR symbol as she plants a tiny green seedling in a little pot. Spray-painted words read: 'From this moment despair ends and tactics begin.' Hallelujah to that.

55 Tufton Street

Some months later, I was invited to speak at a Writers Rebel event with Mark Rylance and Juliet Stevenson. It was a chilly wet evening in London. My partner and I had split up. I was nearly too unhappy to speak, and almost too cold to hold the microphone. But we have to speak: we writers need to be in service to something greater than ourselves, now as never before. We all as humans have to speak, whatever our role is in the world. As cleaners and teachers and doctors and fixers and builders and mothers and makers: all of us have to speak the truth about the climate crisis to those around us, to everyone we meet, to counter the lies that have been spread, and the false doubts cast.

The event was staged on the street outside 55 Tufton Street, home of the climate-denialist Global Warming Policy Foundation. Climate denialists have no right to undermine climate science because they are ludicrously unqualified to do so. They speak without the necessary expertise. So much so, that I want to give you a little flavour of climate denialists and their academic backgrounds.

At the Global Warming Policy Foundation, we have its founders: Nigel Lawson, whose academic expertise is politics, philosophy and economics, which is a far cry from climate science. The GWPF was co-founded by Benny Peiser, who is a sports anthropologist.

Other denialists include James Delingpole, whose climate-science qualifications are simply a degree in English language and literature; Christopher Booker had a history degree; Christopher 'Lord' Monckton studied classics with a diploma in media studies; Andrew Montford studied chemistry then chartered accountancy. Richard A. North is a specialist in public-sector food-poisoning surveillance.

At the time I spoke, the Global Warming Policy Foundation's website quoted papers from: Ole Humlum, a geographer; Indur Goklany, who has a Ph.D, in electrical engineering; and Paul Homewood, who is a former accountant.

And then we have Matt Ridley, whose academic qualification giving him the right to critique and undermine climate science is this: he has a doctorate on sexual selection in pheasants.

These are voices which should never have been listened to. And this is the voice of someone who cannot speak but can only sing and must be heard. She is an endling.

Letter to an Endling

Dear Esa,

Your name, Esa, means 'the only one' in human language: 'the lonely one'.

After your last heartbeat, a world will be gone – for ever. Not, people think, an important one. Only yours, but your only world.

After your last flight, a freedom will be lost – for ever. Not, people think, an important one. Only yours, but your only freedom.

After your last song, a category of music will be silenced – for ever. Not, people think, an important one. Only yours, but your only song and the only one you ever wanted to hear sung back to you. For while you live, you can sing your female laughing thrush call all you like and no male will ever answer you and you will never know why.

Your story happened because of the slaughter of songbirds. In flocks once, then trapped, traded and caged, sentenced to solitary, forced to sing solo. Bird-sorrow for a status-symbol.

You are a nervous bird. In the photograph I have of you, you look frightened. Your eyes are an orange circle with a black centre, and you don't like being in the eye-line of your keepers. You are easily stressed, and would rather be hidden in deep foliage, tucked in thickets of forests.

You have never wanted to call attention to yourself, except for a mate, but now you have the cachet of true tragedy. Your kind, the rufous-fronted laughing thrush, subspecies *slamatensis*, is named as the world's next most likely known extinction. You, exactly you, Esa, one single individual bird, are the last. Your death will mark its extinction. You, Esa the lonely one, as the last individual of your kind, are an endling.

This is what extinction sounds like: the silencing of song that should have been forever yours.

Forever yours,
Jay.

Regina vs *Me*

Nine months after my arrest, and for the first time in my life, I am on trial, for breaching a 'Section 14' order intended to clear rebels off the streets.

In the dock, I take the oath to tell the truth, the whole truth and nothing but the truth.

Telling the truth because other people haven't.

The government has a binding legal obligation to inform the public, and has failed. The media has told partial truths and untruths, actively misinforming the public, undermining the work of climate scientists who have faced death threats for their tenacious truth-telling.

I speak in my own defence. This is my statement.

I want to thank the court for its time and attention, to thank the police officers for their part in this, to thank the judge. I particularly want to thank the court usher, for her presence and her part in creating an atmosphere of kindness. Thank you.

My name is Jay, and I am a writer. I have written about climate change and the living world for decades.

I have also spent a lot of time in Indigenous communities who are likely to be severely affected by the climate crisis. I have been writer in residence at the Potsdam Institute for Climate Impact Research and worked with the organization Tipping Point to communicate the severity and urgency of the climate crisis to the wider public.

Nothing I or any of us has done has had anything like the necessary effect, nothing in proportion to the horror that is to come and, indeed, is already happening for many people in developing countries. No amount of careful peer-reviewed scientific study has appropriately alerted the public, largely because the two avenues through which this information should have been spread – the government and the media – have signally failed in their duty to sufficiently inform the public and at times the media has actively misinformed the public.

By April last year I had decided to risk breaking the law, for the first time in my life. I respect the principle of law. Indeed, one of the things I most fear about the climate crisis is the widespread and terrifying lawlessness of societal breakdown.

My actions caused some inconvenience, for which I am genuinely sorry. I acted in a wholly peaceful way to draw attention to the current deaths and the impending horror of the climate crisis.

I understand a Section 14 can be imposed if there is

danger of serious disruption to the life of the community. In my view, we were attempting to *protect* community in a broad sense, including those not in the immediate vicinity, including the community of the unborn. This is for the children.

The CPS says that I '*must have known that blocking a highway could have no impact on the alleged threat. In the present case, there is no real nexus between the failure to comply with S14 . . . and any actual threat of death or serious injury from climate change . . . the only intention was to raise publicity for the campaign*'. I would argue that there certainly is a real nexus between the failure to comply with S14 and the evils of the climate crisis. That nexus is the media. The CPS is absolutely correct to say the intention was to raise publicity for the campaign. This is not so much their prosecution as it is my defence.

I include an expert witness statement from John Ashton, who was the British government's climate change ambassador and special representative for three successive Foreign Secretaries. My second expert witness statement is from Professor Justin Lewis, Professor at Cardiff University Research Centre in Journalism and Media. My third expert witness statement is from George Marshall, specialist on climate communications for the IPCC and the UK government.

The climate crisis inflicts inevitable and irreparable evil. It is killing people now and will kill far more, particularly children. The World Health Organization says, 'Climatic changes already are estimated to cause over 150,000 deaths annually,' while the report, commissioned by twenty countries from DARA (an independent international organization assessing the impact of humanitarian aid), says, 'Climate Change causes 400,000 deaths on average each year . . . combined climate-carbon crisis is estimated to claim 100 million lives between now and the end of the next decade.' DARA reports that 'the deaths are mainly due to hunger and communicable diseases that affect above all children in developing countries'.

Avoiding a climate crisis cannot be done without first massively increasing public awareness, but neither governments nor the media are adequately informing the public as to the scale, speed and horror of the situation. In the words of Professor Lewis: 'Over the last 30 years media coverage has not matched the significance of the issue or the weight of scientific evidence.' The media has the power to set the agenda and shape public priorities. However, media coverage of protest is low when there is no disruption. Quoting a myriad of academic studies, Professor Lewis shows: 'Peaceful demonstrations with no conflict are less likely to get media attention than protests which involve conflict.' Decades of lawful marches have not worked, but media coverage increases dramatically when there is a willingness to test the law and mass arrests. Professor Lewis writes: 'Media coverage of demonstrations generally increases if the police make arrests. For this reason, Extinction Rebellion have been conspicuously successful in putting climate change on the media agenda.'

XR gained massive coverage and created an urgency in public discourse, significantly altering the attitude of the media towards the climate crisis. Professor Lewis writes: 'It is only over the last year or so that we have finally begun to see a shift towards more serious, significant and sustained media coverage of anthropogenic climate change. This has been in direct response to the School Climate Strikes and coverage of acts of civil disobedience by Extinction Rebellion.'

Widespread press coverage in turn altered public awareness: an Ipsos MORI opinion poll in August 2019 after XR's April Rebellion showed that 85 per cent of Britons are now concerned about climate change, with the majority (52 per cent) very concerned. Citing the effect of Extinction Rebellion, Ipsos MORI's poll showed the highest levels of public concern for climate change in the last fifteen years.

That public awareness led directly to policy change, and a climate emergency was declared by the UK parliament in May 2019.

George Marshall, communications specialist and founder of Climate Outreach, external communications advisers to governments, to the IPCC and the World Bank, shows that the British government has a binding legal obligation to develop educational public awareness on climate change and its effects, and he writes: 'As a specialist in communications, and advisor to many governments, I have no hesitation in saying that the British government has failed to meet these obligations and still has no adequate strategy for achieving them. In my view the effectiveness of Extinction Rebellion in bringing the urgency of climate change to public attention, and the repeated demands of the protesters to "Tell The Truth" are entirely within the spirit of this international commitment. The failure of the government to build a broad-based public mandate for action has required this form of disruptive action.'

I want the court to know that for me it has been incredibly hard – frightening, actually – to get to the point where I break the law and risk becoming a convicted criminal. I have needed a lot of help and support along the way, and I am really touched that, unprompted, my expert witnesses offered the following words. It meant a lot to me to read them.

Former Ambassador John Ashton writes: 'The decision to participate in direct action of any kind is intensely personal, especially if it might lead to arrest and a criminal record. In the case of the climate crisis, future generations will look back on those who made that commitment, and who acted peacefully and with compassion towards their fellow citizens, as heroes in a justified struggle . . . They are far from being criminals.

And if, from an excessively narrow perspective, they can be alleged to have committed any crime, they will be seen to have done so only in pursuit of the true public interest in preventing a greater crime from being perpetrated.'

Professor Lewis writes: 'To criminalise Extinction Rebellion would be, in effect, to penalise an example of good and thoughtful citizenship. It would, in effect, send a message that the minor inconveniences caused by their actions are more important than the most serious threat to life on earth humankind has ever faced, one that places us and future generations at risk . . . History will judge them in the same way we now judge the civil rights movement.'

Sometimes it falls upon a generation to be great, said Nelson Mandela. History is calling from the future, a hundred years from now. Half a hundred years. Ten. Today. Calling the conscience of all of humanity to act with the fierce urgency of now. This is the time. Wherever we are standing is the place. We have just this one flickering instant to hold the winds of worlds in our hands, to vouchsafe the future. This is what destiny feels like. We all need to live by a credo that matters. This is mine. We have to be greater than we have ever been, dedicated, selfless, self-sacrificial.

Humanity itself is on the brink of the abyss: our potential extinction. We face a breakdown of all life, the tragedy of tragedies: the unhallowed horror. Time is broken and buckled, and seasons are out of step so even the plants are confused. Ancient wisdoms are being betrayed: to every thing there *was* a season, a time to be born and a time to be a child, protected and cared for, but the young are facing a world of chaos and harrowing cruelty. We are nature and it is us, and the extinction of the living world is our suicide. Not one sparrow can now be beneath notice, not one bee.

Only when it is dark enough can you see the stars, and they are lining up now to write *rebellion* across the skies. There is no choice. This is a rebellion for the young people and for the grandmothers. This is for the turtle and the salamander, the dugong and the dove. It is for the finned, furry and feathered ones, the ones who scamper and swim, the chattering, chirping and hooting ones.

Each generation is given two things: one is the gift of the world, and the other is the duty of keeping it safe for those to come. This contract is broken, and it is happening on our watch.

The world's resources are being seized faster than the natural world can replenish them. The climate crisis means the future will pay dearly for the actions of the present. Children can do the maths on this, and know they are being sent the bill. And the young are in rebellion now. Why? Because they are the touchstone of nations, carrying the moral authority of innocence. Because they are young enough to know cheating is wrong and old enough to see they have been cheated of their safety, their dreams and their future. Because they are young enough to be awed by the magic of living creatures and old enough to be heartbroken by their slaughter. Because they are young enough to know it is wrong to lie and old enough to use the right words: *this is an emergency*.

Worldwide, the heaviest emissions have been produced by the richest nations, while the heaviest consequences are being felt by the poorest. The few have sown the wind and are forcing the many to reap the whirlwind.

Extinction Rebellion's vision is a politics of kindness. Its vision depends on values that are the most ordinary and therefore the most precious: honesty, decency, fairness and care.

This vision has a map. It is the map of the human heart. Believing in unflinching truth, reckless beauty and audacious love, knowing that life is worth more than money and that there is nothing greater, nothing more important, nothing more sacred than protecting the spirit deep within all life. XR has a rallying cry: *This is life in rebellion for life*. This credo is what brought me to standing in the dock here.

I am wearing this bracelet. A mother came up to me in Oxford Circus with her daughter, about five, who had made this ribboned wristband, and her mother asked me if I would mind wearing it. Would I mind? It'd be an honour. It was to say thank you to people who were willing to be arrested, she said. So I wore this for the child, as she had made it for me. Without knowing each other. And that is the point. That we will hold the hand of strangers, a little bit of me stays with her, and a little bit of her stays with me. The bonds between us all.

I am not going to talk about the law, and I am not going to discuss the fact that the chief inspector has been shown in two cases involving XR in December to have misunderstood the European Human Rights Act. I am not going to talk more about the precise legal grounds for the defence of necessity. I am not going to include all the legal arguments for our right to protest and whether this is being infringed.

I am not going to talk about the law. I am going to talk about justice. At the Rebellion in April last year I saw one of the most poignant things I have ever seen. On Waterloo Bridge, amongst reams of forget-me-nots, a little girl in a red cape was writing in chalk on the tarmac. She wrote: 'There is no planet B so we're asking for your hep.' She examines what she has written, then carefully adds the missing 'l'. Help. She walks away. Then she stops, turns back, studies it again, kneels down and adds 'Please'. It is unbearably painful to see a child on her knees pleading for her life. This is injustice of the most damning sort: that children are facing lives dramatically and horribly affected by the climate crisis. Really seeing the truth of this situation is terrifying and full of grief.

There would be justice if her plea was heard. Justice if her life was protected. It would be fair and just if the next generations could grow up without feeling terrified, with reason, of what the future holds in a world of climate crisis, whose imminent threat of evil will be so great that it holds all other evils within it. All forms of brutality, rape and murder. I feel passionately that we need to protect life – in all its forms – and to act with justice towards future generations.

The job of a writer is to be a good messenger for others, even if it is difficult. A few years ago, I went to West Papua, to talk with Indigenous people there, who are the victims of a genocide because their lands are being seized for extractive industries. It is a microcosm of the situation many Indigenous people are in. A few years before I visited, writers and reporters had been shot for covering this genocide. Yes, I was scared. Yes, I went. It is Indigenous people who are at the sharp end of the climate crisis, they who are losing their land and water, their lives, and their languages and culture. It is their children more than any who risk losing their lives because of the climate crisis. This is for the children.

I love children and always wanted a child, but I am truly relieved now that I do not have one, because I would at some point plead with them not to have a child of their own. My own grandmother was a Pankhurst, and I am happy to follow in those footsteps, except for one thing: I would not want to be a grandmother myself.

I am here looking for justice in all its beautiful forms, knowing that every one of us alive on Earth today needs to be looking to a very different court in a very different future, to assess our guilt or innocence.

In Oxford Circus, I was passed a little handwritten note. It said, 'I can't get arrested because I am only ten but thank you for doing this for me.' It is my vow to live guided by justice for the world I love, a world where in the eyes of a child I am innocent.

In the dock, I felt desolate with isolation. I couldn't read the judge's face. In the middle of my statement, I suddenly thought I was going to cry, and the judge asked if I needed to take a break. I went on to the end. Only then could I see the judge's expression. His eyes were red and looked full of tears. Then the verdict. 'It is with a really heavy heart that I have to convict you.'

He ruled as a judge, then he spoke as a man. He expressed his belief that 'we have a responsibility to the whole of the world,' adding, 'I accept and believe that with all my being.' And then, all eyes on him, Judge Noble, this judge in a mad situation, nobility of soul at odds with the circumstances of the law, said: 'This is going to be my last Extinction Rebellion trial for a little while. I think they only allow us to do so many of these before our sympathies start to overwhelm us. When I started, I was fully expecting to see the usual crowd of anarchists and communists, and all the dreadful things the Daily Mail *say you are. I have to say I have been totally overwhelmed by all the defendants. It is such a pleasure to deal with people so different from all of the people I deal with in my regular life. Thank you for your courtesy, thank you for your integrity, thank you for your honesty. You have to succeed.'*

And that is when everything changed. Sometimes it falls upon a judge to be great, who used that moment to speak truth to the power that is the Daily Mail *and bestowed on us all a generous and chivalrous tribute. 'To speak the true word is to transform the world,' said educationalist Paolo Freire. As the court rose, we were, all of us, standing in a place of grace. Because that is where – and why – the real truths are told.*

Thanks

The author wishes to thank the estate of Leonard Cohen for kind permission to quote from 'Hallelujah' and 'Bird on a Wire'.

'The Solar Flares of Fascism' and 'The Forests of the Mind' first appeared in *Aeon* magazine <aeon.co>

'Happiness, Animals and the Honeyguide Rule' first appeared in *Lapham's Quarterly* <laphamsquarterly.org>

'Coral's Swan Song' and 'This England' first appeared in *Dark Mountain* <dark-mountain.net>

A version of 'Let There be Dark' first appeared in *Emergence* magazine <emergencemagazine.org>

'A Time of Rebellion' first appeared in *Orion* magazine <orionmagazine.org>

'Letter to an Endling' first appeared in *Letters to the Earth*, edited by Culture Declares Emergency

Kind thanks to all.

*

Thanks

With thanks to dear friends along the way: Jan, Euan, Marg, Andy, Tuppin, Hannah, Ralph, Rachel, George, Anna, Giuliana, Rob, Buz, Thoby, Niall, Deborah, Mike, Preds, Vic, Nicoletta, Jules, Philip, Thea, Maia, Anita, Andy, Clare, Alice, John, Mel, Catrina, Gareth, Angela, Sara, Boff, Casey, Jenny, David, Timothy, John, Jan, Jen, David, Tom, Ben, Iain, Henry, Pippa and Mark.

To my most excellent editor Simon Prosser and my peerless agent Jessica Woollard, heartfelt appreciation and glad gratitude.

Dedication

This book is dedicated to my mother,
who gave me the green world and the words,
with all my love